Concrete Masonry Handbook

for Architects, Engineers, Builders

Concrete Masonry Handbook

for Architects, Engineers, Builders

by W.C. Panarese, S.H. Kosmatka, and F.A. Randall, Jr.

PORTLAND CEMENT **PCA** ASSOCIATION

5420 Old Orchard Road, Skokie, Illinois 60077-1083
Phone: 708/966-6200 • Facsimile: 708/966-9781

An organization of cement manufacturers to improve and
extend the uses of portland cement and concrete through
market development, engineering, research, education, and
public affairs work.

About the authors: The engineers responsible for the final manuscript are William C. Panarese and Steven H. Kosmatka, manager and assistant manager, respectively, Construction Information Services, Portland Cement Association, and Frank A. Randall, Jr., consulting structural engineer (formerly with PCA, now retired).

On the Cover: Loadbearing cavity wall combines split-ribbed and regular concrete masonry units in the new Congregation B'Nai Jeshurun Barnert Temple, Franklin Lakes, New Jersey. Architect: Percival Goodman, FAIA; Project Architect: Wells Associates; General Contractor: Visbeen Construction Co.; Photographer: Bo Parker.

This publication is based on the facts, tests, and authorities stated herein. It is intended for the use of professional personnel competent to evaluate the significance and limitations of the reported findings and who will accept responsibility for the application of the material it contains. The Portland Cement Association disclaims any and all responsibility for application of the stated principles or for the accuracy of any of the sources other than work performed or information developed by the Association.

Printed in the United States of America

Fifth Edition print history

First printing 1991

Library of Congress Cataloging-in-Publication Data

Panarese, William C.
    Concrete masonry handbook for architects, engineers, builders/by W.C. Panarese, S.H. Kosmatka, and F.A. Randall, Jr.—5th ed.
        p.   cm.
    Includes bibliographical references (p. 219) and index.
    ISBN 0-89312-093-6 (pbk.)
    1. Concrete construction—Handbooks, manuals, etc.
2. Concrete masonry—Handbooks, manuals, etc. I. Kosmatka, Steven H. II. Randall, Frank Alfred, 1918-. III. Title.
TH1461.P34   1991
693'.5—dc20   90-64060

Caution: Avoid prolonged contact between unhardened (wet) cement or concrete mixtures and skin surfaces. To prevent such contact, it is advisable to wear protective clothing. Skin areas that have been exposed to wet cement or concrete, either directly or through saturated clothing, should be thoroughly washed with water.

EB008.05M

ACKNOWLEDGMENTS

Many individuals and organizations within the concrete masonry industry have assisted in writing this expanded Fifth Edition of the *Concrete Masonry Handbook*. Trade associations, professional societies, manufacturers, architects, contractors, consultants, and many others have provided research material and illustrations as well as manuscript reviews and commentary. We are particularly grateful for the guidance and fine cooperation of the individuals listed below. This acknowledgment does not necessarily imply approval of the text by these individuals since the entire handbook was not reviewed by all those listed and since final editorial prerogatives have necessarily been exercised by the Portland Cement Association.

Mario J. Catani, President, Dur-o-Wal, Inc.

Albert W. Isberner, Consulting Materials Engineer

Eduardo A.B. Salse, President, Salse Engineering Associates, Inc.

The following staff of the Portland Cement Association (PCA) and Construction Technology Laboratories, Inc. (CTL) provided valuable technical assistance, commentary, and manuscript review:

James P. Hurst, senior fire protection engineer, Engineered Structures and Codes, PCA

John M. Melander, masonry specialist, Engineered Structures and Codes, PCA

Jake W. Ribar, principal masonry evaluation engineer, Structural Engineering, CTL

Publications of the following organizations are frequently cited:

American Concrete Institute (ACI)

American Society for Testing and Materials (ASTM)

National Concrete Masonry Association (NCMA)

PREFACE

Concrete masonry is a versatile, durable, and economical construction material. Since 1882 when the first concrete block was molded, concrete masonry has become a staple building material in the construction industry. Concrete masonry is used in residential, commercial, and industrial structures, plus many special applications such as paving, sound barrier walls, and retaining walls. It is used to support structural loads and provide esthetic appeal.

The *Concrete Masonry Handbook* has been the primary reference on concrete masonry since it was first published in 1951. As with past editions that reflected progress and advances in the masonry industry, this fifth edition also presents new developments and engineering concepts that have come of age since the fourth edition was published in 1976.

The layout of the book is similar to that of the previous edition, except for a new chapter on maintenance of masonry structures. A major emphasis has been placed on requirements of the new *Building Code Requirements for Masonry Structures* (ACI 530/ASCE 5) and the *Specifications for Masonry Structures* (ACI 530.1/ASCE 6). New ASTM standards are also frequently cited and an extensive list of ASTM standards used in the masonry industry is included in the References.

This fifth edition is 36 pages longer than the fourth. Along with updating of all previous material, a number of new topics have been added including tuckpointing, cleaning masonry, foundation block, ready mixed mortar, modified mortar, silo-mixed mortar, paving unit installation, masonry veneer over steel stud wall systems, exterior insulation and finish systems, and in the Appendix, new flashing details. Most of the new or expanded material is accompanied by explanatory photographs, figures, or tables.

The authors are grateful for contributions made to the fifth edition by leading experts in the concrete masonry field (see *Acknowledgments*). The authors have tried to make this edition a concise and current reference on concrete masonry technology. As there is always room for improvement, readers are encouraged to submit comments to improve future printings and editions of this book.

CONTENTS

Chapter 3.
Properties of Concrete Masonry Walls

Chapter 7.
Applied Finishes for Concrete Masonry 161

Chapter 8.
Various Applications of Concrete Masonry 173

Chapter 9.
Maintenance of Masonry Structures

Fig. 1-1. Concrete masonry units are available in a wide variety of sizes and shapes. *Photo courtesy of the Besser Company.*

Chapter 1

GENERAL INFORMATION ON CONCRETE MASONRY UNITS

Concrete masonry is widely used to construct small and large structures because of its attractive appearance, minimum maintenance, safety, and economy. Concrete masonry provides an effective barrier to sound and reduces internal temperature variations and peak loads on heating and cooling systems. It provides architectural freedom and versatility with striking esthetic appeal. Almost any shape of structure is possible to construct as demonstrated in Figs. 1-2 and 1-3.

Concrete masonry is not easily damaged. It resists weathering and vandalism. Concrete masonry's durability and minimum maintenance extend a building's useful life, providing an enduring, high-quality appearance. Special architectural units require no costly additional architectural treatments.

The concrete masonry units discussed throughout this publication often are referred to as concrete block or concrete brick. Concrete block is usually a relatively large unit with hollow cores. Though sizes vary, a typical concrete block is the rectangular 8×8×16-in. unit. Concrete brick basically matches the size and scale of regular clay brick. It is often impossible to distinguish high-quality concrete brick from clay brick.

Concrete masonry units (block and concrete brick) are available in sizes, shapes, colors, textures, and profiles for practically every conceivable need and convenience in masonry construction. In addition, concrete masonry units may be used to create attractive patterns and designs to produce an almost unlimited range in architectural treatments of wall surfaces. The list of current applications is lengthy, but some of the more prominent uses are for:

- Exterior load-bearing walls (below and above grade)
- Interior load-bearing or non-load-bearing walls
- Fire walls, party walls, curtain walls
- Partitions, panel walls, solar screens
- Backing for brick, stone, stucco, and exterior insulation and finishing systems
- Veneer or nonstructural facing for wood, steel, concrete, or masonry
- Fire protection of structural steel members
- Firesafe enclosures of stairwells, elevator shafts, storage vaults, or fire-hazardous work areas
- Piers, pilasters, columns
- Bond beams, lintels, sills
- Retaining walls, slope protection, ornamental garden walls, and highway sound barriers
- Chimneys and fireplaces (indoor and outdoor)
- Catch basins, manholes, valve vaults
- Paving and turf block

Specifications and Codes

As with every type of building material, there is a multitude of specifications and codes to guide or regulate manufacturers, designers, and builders. To supplement the information in this book, the reader is strongly encouraged to obtain copies of *Building Code Requirements for Masonry Structures, Specifications for Masonry Structures,* and the Commentaries to these documents (see References 75-77 and other prominent references given at the end of this book).

General Requirements

Concrete masonry units are manufactured in the United States to conform to requirements of the American Society for Testing and Materials (ASTM). ASTM specifications classify concrete masonry units according to grade and type. The grade describes the intended use of the concrete masonry units, as shown in Table

Fig. 1-2. Triangular shaped, split-faced block structure.

Fig. 1-3. Circular structure using block backup.

Table 1-1. Grades of Concrete Masonry Units for Various Uses*

Grade	ASTM C90 or C145 (block)	ASTM C55 (concrete brick)**
N	For general use such as in exterior walls below and above grade that may or may not be exposed to moisture penetration or the weather and for interior walls and backup.	For use as architectural veneer and facing units in exterior walls and for use where high strength and resistance to moisture penetration and severe frost action are desired.
S	Limited to use above grade in exterior walls with weather-protective coatings and in walls not exposed to the weather.	For general use where moderate strength and resistance to frost action and moisture penetration are required.

*Adapted from ASTM C55, Standard Specification for Concrete Building Brick; ASTM C90, Standard Specification for Hollow Load-Bearing Concrete Masonry Units; and ASTM C145, Standard Specification for Solid Load-Bearing Concrete Masonry Units. Not applicable to ASTM C129, Standard Specification for Non-Load-Bearing Concrete Masonry Units.

**Also applicable to solid concrete veneer and facing units larger than brick size, such as split block.

1-1, while the type of unit is indicated as either Type I, moisture-controlled, or Type II, non-moisture-controlled. Type I units are specified where drying shrinkage of the block due to loss of moisture could result in excessive stress and cracking of walls.

In common with a number of building materials, concrete shrinks slightly with loss of moisture. The amount of moisture loss is affected by the relative humidity of the surrounding air. After concrete has dried to constant moisture content at one atmospheric condition, a decrease in humidity causes it to lose moisture or an increase causes it to gain. When moist units are placed in a wall and this inherent shrinkage is restrained, as it often is, tensile and shearing stresses are developed that may cause cracks in the walls.

Where concrete block or concrete brick walls will be exposed to low relative humidities, and in areas of the country having exceptionally dry climates, the maximum moisture content permitted at the time of delivery should be lower than when the construction is located in a more humid environment. Furthermore, concrete masonry units have varying inherent drying shrinkage characteristics depending on method of manufacture and materials used. In an attempt to equalize the drying shrinkage, units with low shrinkage characteristics are allowed to have higher moisture contents than units with high shrinkage. Requirements for moisture-controlled units are shown in Table 1-2.

Compressive strength is an important property of concrete masonry and related to the general usage (described in Table 1-1). Strength requirements are given in Table 1-3.

To understand the significance of the different ASTM strength requirements, it is important to note that net area and shape play an important role. First, the concrete strengths for the respective grades of solid and hollow block or for concrete brick are essentially the same. However, a "solid" block with 75% concrete has more material to resist load than a hollow block with about 50% solid material. Also, a flat concrete brick will have a considerably higher crushing strength than

Table 1-2. Moisture Content Requirments for Type I Units*

Linear shrinkage, percent**	Maximum moisture content, percent, based on total absorption (average of 3 units)†		
	Humid	Intermediate	Arid
0.03 or less	45	40	35
From 0.03 to 0.045	40	35	30
0.045 to 0.065 (max.)	35	30	25

*Adapted from ASTM C55, C90, C129, and C145.

**Based upon ASTM Method C426, Standard Method of Test for Drying Shrinkage of Concrete Block, conducted not more than 12 months prior to delivery of units.

†Consider the following humidity conditions at jobsite or point of use: humid, average annual relative humidity above 75%; intermediate, average annual relative humidity 50-75%; and arid, average annual relative humidity less than 50%. (Mean annual relative humidities for the United States are shown in Fig. 6-71).

Note: Moisture controlled requirements are difficult to understand. The intent of the table is to have the manufacturer qualify his particular mixtures by having linear shrinkage tests performed and then that mixture, when used in concrete masonry units, can be later tested for moisture content to allow conformance with the product specification. It is near impossible to establish conformance of the product to the specification after construction. In the state of Wisconsin, moisture controlled units are not produced. Non-moisture controlled units are being used successfully by being attentive to joint reinforcement and spacing of control joints.

Table 1-3. Strength and Absorption Requirements for Concrete Masonry Units

Type of unit	ASTM designation	Grade of unit	Minimum compressive strength, psi, on average gross area		Maximum water absorption, pcf, based on oven-dry unit weight			
			Average of three units	Individual unit	Lightweight concrete		Medium-weight concrete, 105 to 125 pcf	Normal-weight concrete, 125 pcf or more
					Less than 85 pcf	Less than 105 pcf		
Concrete brick	C55	N	3,500*	3,000*	—	15	13	10
		S	2,500*	2,000*	—	18	15	13
Solid load-bearing block	C145	N	1,800	1,500	—	18	15	13
		S	1,200	1,000	20	—	—	—
Hollow load-bearing block	C90	N	1,000	800	—	18	15	13
		S	700	600	20	—	—	—
Hollow and solid non-load-bearing block	C129	—	600**	500**	—	—	—	—

*Concrete brick tested flatwise.
**Minimum compressive strength, psi, on average net area.

the shell of a hollow block, just as a 2¼-in.-high column 4 in. wide will be stronger than an 8-in.-high column 1¼ in. wide.

The economic use of concrete masonry over a wide range of applications is made possible by availability of units with a wide range of strengths. Units called "high-strength" block are gaining in popularity, especially in some areas. Their manufacture is slightly different from regular units; they have to be produced more slowly and the selection of block mix must be made more carefully.

High-strength block are not yet defined by a national specification, but they have been considered to fall in the following strength classes.

	Gross area strength, psi	Net area strength (53% solid units), psi
Regular-strength block	1,060	2,000
High-strength block	1,860	3,500
Extra-high-strength block	2,650	5,000

The high-strength block can be manufactured with most aggregates or combinations of aggregates by all block producers. The extra-high-strength block is usually limited to applications where it is required to limit wall thickness in buildings over 10 stores high. Drying shrinkage of high-strength units may be twice that of units made under ASTM C90. On the other hand,

absorption decreases with strength and density of the units.*

Density is related to the amount of water absorption of concrete masonry units as shown in Table 1-3. These properties affect construction, insulation, acoustics, appearance, porosity, painting, etc. Absorption affects the quality of mortar needed. If a masonry unit absorbs water too fast, the mortar will need more water retentivity. This is necesary to give the mason time to place and adjust the block before the mortar stiffens, and to achieve a strong mortar bond. Absorption rate is a specification requirement for clay units but not for concrete masonry. However, the principle applies and control is exercised by limiting the amount of absorption.

Architectural concrete masonry units for interior use should comply with the requirements of ASTM C90 and C145 for hollow and solid load-bearing units, respectively. Architectural concrete masonry units for exterior use are often specified to conform with specifications for concrete building brick, ASTM C55, Type N.

The ASTM specifications for concrete masonry do not fix the weight, color, surface texture, fire resistance, thermal transmission, or acoustical properties of the units.

---

*See Ref. 44.

Manufacture

Concrete masonry units are made mainly of portland cement, graded aggregates, and water. Depending upon specific requirements, the concrete mixtures may also contain other suitable ingredients such as an air-entraining agent, coloring pigment, siliceous and pozzolanic materials, and water repellants.

Mass production has contributed to the relatively low cost of quality concrete masonry units. In many production plants some phases of the manufacturing process are completely automated.

Briefly, the manufacturing process involves the machine-molding of very dry, no-slump concrete into the desired shapes, which are then subjected to an accelerated curing procedure. This is generally followed by a storage or drying phase so the moisture content of the units may be reduced to the specified moisture limits prior to shipment. The concrete mixtures must be carefully proportioned and their consistency controlled so that texture, color, dimensional tolerances,

and other desired physical properties are obtained. High-strength units have concrete with higher cement contents and more water, but still have no slump. Automatic machines consolidate and compact these concretes by vibration and pressure, and mold approximately a thousand 8×8×16-in. masonry units (or their equivalent in other sizes) per hour.

Two types of accelerated curing are utilized by the concrete masonry industry, with variations according to local plant requirements and raw materials used. The more common type of curing provides for heating the block in a steam kiln at atmospheric pressure to temperatures ranging from 120 to 180 deg. F. for periods up to 18 hours.* Atmospheric pressure methods may require subsequent accelerated drying treatment or a period of natural drying in the storage yard under protective cover. In a variation of low-pressure curing, a carbonation stage is added to reduce the shrinkage characteristics of the masonry units.

---

*See Ref. 70.

Fig. 1-4. A modern block plant in operation. In the left background is the concrete mixer. Concrete travels through the large hopper to the block machine. At the right is a curing rack unloading block to a conveyor.

Fig. 1-5. Loading block into autoclave for curing with high-pressure steam.

Although rarely used, the other type of curing is known as autoclaving or high-pressure steam curing (Fig. 1-5).* In this process the units may be subjected to saturated steam at 325 to 375 deg. F. maximum temperatures (80 to 170 psig) in a large cylindrical pressure vessel for various times up to 12 hours. The interval at maximum temperature is preceded by a preset period of 1 to 4 hours, followed by pressure-buildup time of about 3 hours. At the end of the steaming interval it is common practice to drain the condensate and reduce the cylinder pressure as rapidly as possible—within 20 to 30 minutes. By this proce-dure, stored heat quickly lowers the absorbed moisture content in the units to specification requirements.

For storage or shipment to the building site the units are generally placed in small stacks or "cubes" consist-ing of layers of fifteen to eighteen 8×8×16-in. units per layer, or the equivalent volume in other sizes. The cubes (40×48×48 in. or 48×48×48 in.) may be assem-bled on wooden pallets or banded with the bottom layer of block having cores positioned horizontally. Most delivery trucks are equipped with a device for unloading the cubes at storage areas on the building site. The number of units per pallet varies with the size of the unit; a pallet of 8×8×16-in. block will have about 100 units while a pallet of regular brick has about 500 units.

Types

Available to designers are units having a wide selection of weights, sizes, shapes, and exposed surface treat-ments for virtually any architectural and/or structural function. Manufacturers and their local associations can supply literature that describe available types of units. *In advance of any detailing the designer is urged to determine the sizes, shapes, textures, and other properties in masonry units required for the proposed construction as well as their availability from local producers.***

Normal-Weight and Lightweight Masonry Units

The terms "dense or normal-weight" and "lightweight" are derived from the density of the aggregates used in the manufacturing process. The normal-weight aggre-gates used are sand and gravel, crushed stone, and air-cooled blast-furnace slag. The lightweight aggregates include expanded shale, clay, and slate; expanded blast-furnace slag; sintered fly ash; coal cinders; and natural lightweight materials such as pumice and scoria. In 1988-89, the United States saw the debut of a cold-bonded, pelletized-fly ash aggregate (agglomerated) consisting of fly ash, cement or lime, and admixtures. This new fly ash aggregate (Fig. 1-6) can be used in lightweight or normal-weight masonry units. In gener-al, local availability determines the use of any one type of aggregate. In some locations the term "concrete block" has been used to designate only those units made with sand and gravel or crushed stone aggregates.

Fig. 1-6. Pelletized aggregate of fly ash and cement.

*See Ref. 2.

**To obtain a list of local producers, consult the yellow pages of the telephone directory and contact the National Concrete Masonry Association, 2303 Horse Pen Road, P.O. Box 781, Herndon, Virginia, 22070-0781, Phone: 703-435-4900, Fax: 703-435-9480.

Table 1-4. Concrete Unit Weight Ranges with Various Aggregates

Concrete	Unit weight, pcf
Sand and gravel concrete	130-145
Crushed stone and sand concrete	120-140
Air-cooled slag concrete	100-125
Coal cinder concrete	80-105
Expanded slag concrete	80-105
Pelletized-fly ash concrete	75-125
Scoria concrete	75-100
Expanded clay, shale, slate, and sintered fly ash concretes	75-90
Pumice concrete	60-85
Cellular concrete	25-44

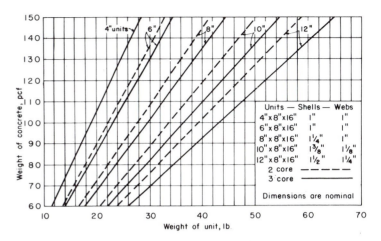

Fig. 1-7. Weights of some hollow units for various concrete densities.

Generally speaking, however, concrete block are made with any of the above aggregates.

The weight class of a concrete masonry unit is based upon the density or oven-dry weight per cubic foot of the concrete it contains. A unit is considered as lightweight if it has a density or unit weight of 105 pcf or less, medium-weight if it has a density between 105 and 125 pcf, and normal-weight if it has a density of more than 125 pcf. Concretes containing various aggregates range in unit weight as shown in Table 1-4.

In addition to the concrete density, the weight of an individual concrete masonry unit depends upon the volume of concrete in the unit. The design of the unit in turn affects its volume. Approximate weights of various hollow masonry units may be determined from Fig. 1-7. Use of lightweight aggregates can reduce the weight about 20 to 45%, compared with the weight of similar units made of normal-weight aggregates, without sacrifice of structural properties. Whether the designer or builder chooses lightweight or normal-weight units generally depends upon availability and the requirements of the structure. Another consideration is that worker efficiency is better with lightweight block than normal-weight block

Hollow and Solid Units

Concrete block are classified as hollow or solid units. A hollow unit is defined as one in which the net concrete cross-sectional area parallel to the bearing face is less than 75% of the gross cross-sectional area (ASTM C90 and C129). Units having net concrete cross-sectional areas of 75% or more are classified as solid units (ASTM C129 and C145*). The net concrete cross-sectional areas of most concrete masonry units range from 50 to 70% (30 to 50% core area) depending on unit width, face-shell and web thicknesses, and core

Fig. 1-8. Low-rise residential building of load-bearing concrete masonry.

---

*ASTM C145, Specification for Solid Load-Bearing Concrete Masonry Units, was withdrawn in 1990 due to a lack of use of this product. It is mentioned here for reference purposes.

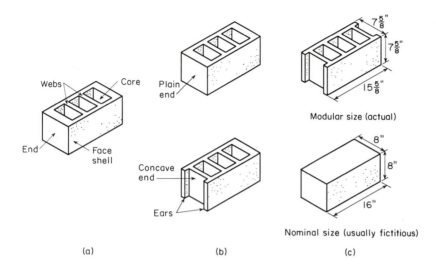

Fig. 1-9. (a) Parts of hollow block. Each block has two face shells, two ends, and one or more webs. (b) Common end types. Special ends are illustrated in Figs. 1-10 through 1-12. (c) Block size nomenclature.

configuration. Because of their reduced weight and ease of handling, hollow units are in greater use than solid units. Fig. 1-9 illustrates the various parts of hollow block.

For structural reasons some standards governing concrete masonry units stipulate minimum face-shell and web thicknesses as shown in Table 1-5. ASTM non-load-bearing hollow units (ASTM C129) are not subject to these requirements, except that the face shell cannot be less than ½ in.

Solid masonry units are used only for special needs, such as for structures having higher than usual design stresses, for the top or bearing course of load-bearing walls, for increased fire protection, or for catch basin or manhole construction. Concrete brick and some split-block units are made 100% solid (without cores), although some concrete brick designs include a shallow depression, often called a "frog," in one bearing face (Fig. 1-12d). However, the net volume of the frog should be not more than 15% of the gross volume of the unit. The purpose of the frog is to reduce weight, provide for better mechanical bond, and prevent the unit from floating when laid in the wall.

Modular Sizes

Concrete masonry unit dimensions are for the most part based on some module, usually 4 or 8 in. From common usage the ⅜-in.-thick mortar joint has become standard. Accordingly, the exterior dimensions of modular units are reduced by the thickness of one mortar joint, ⅜ in. Thus, when laid in a wall the

Table 1-5. Minimum Thickness of Face Shell and Webs*

Nominal width of unit, in.	Minimum face-shell thickness, in.**	Web thickness	
		Minimum web, in.**	Minimum equivalent web, in./lin.ft.†
3 and 4	¾	¾	1⅝
6	1	1	2¼
8	1¼	1	2¼
10	1⅜ (1¼)††	1⅛ (1⅛)	2½ (2½)
12	1½ (1¼)††	1⅛ (1⅛)	2½ (2½)

*Adapted from ASTM C90.
**Average of measurements on three units taken at the thinnest point.
†Sum of measured thickness of all webs in a unit times 12 divided by length of unit.
††This face-shell thickness is applicable where the allowable design load is reduced in proportion to the reduction in thickness from the basic face-shell thickness shown.

modular units produce wall lengths, heights, and thicknesses that are multiples of the given module. This permits the designer to plan building dimensions and wall openings that will minimize the expense of cutting units on the job.

It is common practice in specifying concrete block to give the wall or block width first, the course (or block) height second, and the block length third, followed by the name of the unit. Dimensions given are nominal, with actual unit dimensions being ⅜ in. less.

Fig. 1-9c shows the dimensional details for an 8×8×16-in. unit, the nominal block size that dominates the industry.

Block is commonly available in widths of 2, 4, 6, 8, 10, and 12 in. Typical nominal block heights are 4 and 8 in., but 12-in. units are also available. Half-length units (nominally 8 in.) and low or half-height units are made as companion items for the mason's convenience in completing various patterns. In some areas the 4-in. module is popular with nominal unit lengths of 8, 12, 16, 20, and 24 in.

Some manufacturers may make units in full nominal dimensions or in stated dimensions other than modular. Whether units are modular or nonmodular, ASTM standards require that tolerances from the manufacturer's catalogued dimensions not exceed ±⅛ in. The control in block plants is such, however, that actual variations are seldom greater than ±1/16 in. The width of split-block units should not be specified within these tolerances. Uniformity of units results in uniform mortar-joint thickness and a pleasing appearance.

Core Types

There are many masonry unit design variations. For example, some manufacturers make either two- or three-core units exclusively, while others make some sizes and shapes in both core designs, with the balance of their production in only one or the other core design.

Fig. 1-10 shows some of the variations in core and end designs as applied to 8×8×16-in. concrete masonry units. Some producers make regular stretcher units with flanged ends in the 8-, 10-, and 12-in. widths. Others have adopted, for regular production, single and double plain-ended designs, thereby meeting needs for stretcher, corner, or pier units from single stocks of a given width. In general, all 4-in. and most 6-in.-wide hollow units are made with plain ends but may contain either two or three cores.

The cores of hollow units are tapered to facilitate stripping in the machine molding process. Some core designs also include a degree of flaring of the face or web to give a broader base for mortar bedding and for better gripping by the mason. The face shells are sometimes thickened as indicated in Fig. 1-10c; this is done to provide greater tensile strengths at these regions in the finished wall.* End flanges may be grooved or plain.

Size and Shape Variations

Fig. 1-10 and the remainder of the drawings in this chapter present a sampling of the large number of sizes and shapes of concrete masonry units from which the designer may choose to build economical and attractive concrete masonry structures. A complete listing is not possible here as some plants produce several hundred different items. Some sizes and shapes are limited to certain areas or are made only on special order. Naturally, there are added costs involved in stocking numerous sizes and shapes, especially if demand is limited. This explains why *there may be limited availability in certain areas.*

Fig. 1-10 shows some types of stretcher and corner units employed in conventional concrete masonry wall construction. Fig. 1-11 includes some of the shapes that are available for the builder's convenience for conventional concrete masonry walls or for a particular need. Half-length units are generally available in most of the shapes shown. Alternately, an easily split slotted or kerfed two-core unit may be used (see far right unit in Fig. 1-10a), or a masonry saw may be used to cut special shapes or shorter lengths from whole units.

Corner blocks have one flush end for use in pilaster, pier, or exposed corner construction. Bullnose blocks have one or more small radius-rounded corners and are used instead of square-edged corner units to minimize chipping. Jamb or sash blocks are used to facilitate the installation of windows or other openings. Capping blocks have solid tops for use as a bearing surface in the finishing course of a wall. Header blocks have a recess to receive the header unit in a composite masonry wall. Header units are essentially no longer used in new construction of exterior walls. Return or corner-angle blocks are used in 6-, 10-, and 12-in.-thick walls at corners to maintain horizontal coursing with the appearance of full-length and half-length units.

Fig. 1-12a shows a number of block types available for partition construction and facing unit backup. Solid units and cap or paving units (Fig. 1-12b and c)** are manufactured in a variety of sizes for use as capping units for parapet and garden walls and for use in stepping stones, patios, fireplaces, barbecues, or veneer. They may be used both structurally and nonstructurally. When they are used in reinforced walls, the reinforcing steel is generally placed in grout spaces between wythes.

Some typical concrete brick units are shown in Figs. 1-12d, and 1-13 through 1-15. Concrete brick is sized to be laid with ⅜-in. mortar joints, resulting in modules of 4-in. widths and 8-in. lengths. The thickness of mortar joints is increased slightly so that three courses (three brick and three bed joints) lay up 8 in. high. Some manufacturers produce double brick units.

Slump units (Fig. 1-12d and e) are produced by using a concrete mixture finer and wetter than usual. The concrete brick or block unit is squeezed to give a bulging effect. The founded or bulging faces resemble handmade adobe, producing a pleasing appearance (Fig. 1-16).

Split brick or block (Fig. 1-12d, f, g, and h) are solid or hollow units that are fractured (split) lengthwise or crosswise by machine to produce a rough stone-like

---

*See Refs. 3 and 6.
**Other units of this type are shown in Chapter 8.

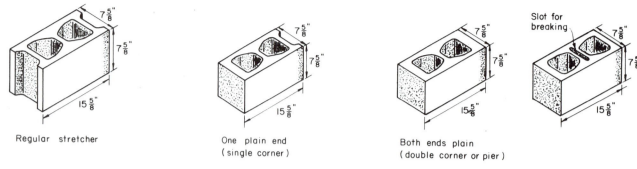

(a) Two-core 8x8x16-in. units

Regular stretcher — One plain end (single corner) — Both ends plain (double corner or pier) — Slot for breaking

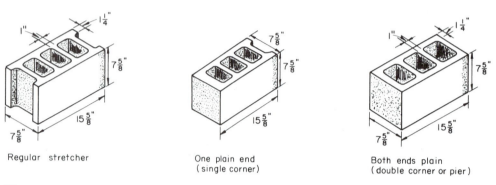

(b) Three-core 8x8x16-in. units

Regular stretcher — One plain end (single corner) — Both ends plain (double corner or pier)

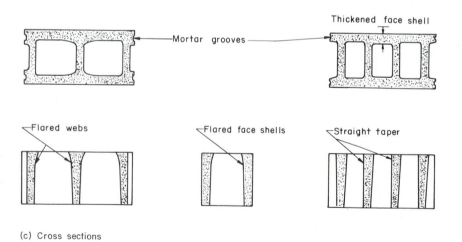

Mortar grooves — Thickened face shell

Flared webs — Flared face shells — Straight taper

(c) Cross sections

Fig. 1-10. Variations in core and end details of 8x8x16-in. units. Core types shown are used also for 4-, 6-, 10-, and 12-in.-wide units. Generally, 4- and 6-in. units are made with both ends plain. *Consult local producers for specific available units.*

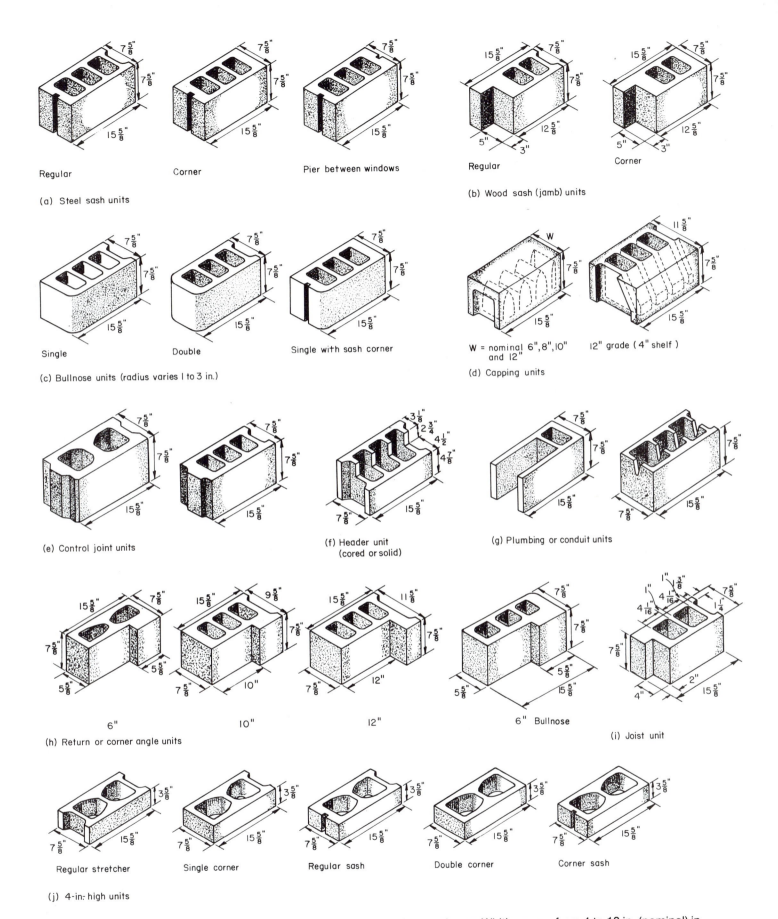

(a) Steel sash units

Regular Corner Pier between windows

(b) Wood sash (jamb) units

Regular Corner

(c) Bullnose units (radius varies 1 to 3 in.)

Single Double Single with sash corner

(d) Capping units

W = nominal 6″, 8″, 10″ and 12″   12″ grade (4″ shelf)

(e) Control joint units

(f) Header unit (cored or solid)

(g) Plumbing or conduit units

(h) Return or corner angle units

6″ 10″ 12″ 6″ Bullnose

(i) Joist unit

(j) 4-in. high units

Regular stretcher Single corner Regular sash Double corner Corner sash

Fig. 1-11. Some sizes and shapes of concrete masonry units for conventional wall construction. Dimensions shown are actual and used for modular construction with ⅜-in. joints. Half-length units are generally available in shapes shown. Widths range from 4 to 12 in. (nominal) in 2-in. increments for some shapes. *Consult local producers for availability before specifying.*

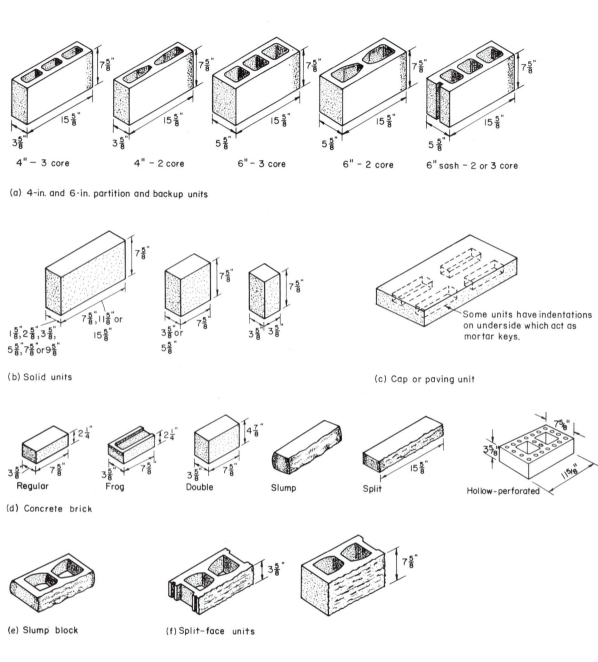

4" – 3 core   4" – 2 core   6" – 3 core   6" – 2 core   6" sash – 2 or 3 core

(a) 4-in. and 6-in. partition and backup units

(b) Solid units

(c) Cap or paving unit

Some units have indentations on underside which act as mortar keys.

Regular   Frog   Double   Slump   Split   Hollow-perforated

(d) Concrete brick

(e) Slump block

(f) Split-face units

Split line

(g) Split block yielding two units.

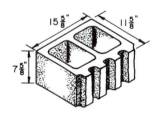

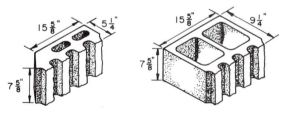

(h) Ribbed split-face units

Fig. 1-12.   Some sizes and shapes of concrete masonry units for partition and backup block, solid and cap or paving block, concrete brick, and slump and split units. *Consult local producers for specific available units.*

Fig. 1-13. Regular sized concrete brick.

Fig. 1-14. Rusticated concrete brick has a rough, rugged texture.

Fig. 1-15. Rusticated concrete brick is popular in residential construction.

Fig. 1-16. Slump-block garden wall.

Fig. 1-17. Exterior wall of 4-in.-high split-face units.

texture (Fig. 1-17). The fractured face or faces, which are exposed when the units are laid, are irregular but sharp, breaking through and exposing the aggregates in the various planes of fracture. By variation of cements, aggregates, color pigments, and unit size, a wide variety of interesting colors, textures, and shapes are produced. The nominal length of split units is 16 in., but

half-length units, return corners, and other multiples of 4 in. are obtainable. The solid units are nominally 4 in. wide and available in various modular heights ranging from $1\frac{5}{8}$ to $7\frac{5}{8}$ in. Split solid units are used as a veneering or facing material. Ribbed hollow units, which can be split to produce unusual effects, are widely used for through-the-wall applications indoors or out.

Fig. 1-18 shows some of the specialized shapes for constructing window sills, copings, bond beams, and lintels. Reinforced bond beams are useful to minimize cracking due to temperature and moisture change (Figs. 1-19 and 1-20). In areas of recurring earthquakes or hurricanes, where major damage to construction has a high probability of occurrence, reinforced bond beams as well as vertical and horizontal wall reinforcement are mandatory. Reinforced lintels are necessary to bridge over openings for windows and doors. The various sizes and shapes of lintels shown are for different requirements of load capacity, spans, wall widths, and window or door types.

Figs. 1-21 and 1-22 illustrate some types of units made for the construction of pilasters, columns, and chimneys.

Fig. 1-23 shows some typical masonry unit shapes made for the construction of sewer manholes, catch basins, valve vaults, and other underground structures. Some unit designs include matching tongue and groove ends. In some areas a similar type of concrete masonry unit may be available for use in the construction of silos and similar containers that must resist internal pressures. These units may be equipped with not only matching tongue and groove ends, but also slots in the bed planes for keyed horizontal joints. Depending upon the service requirements, the whole structure might also be hooped with steel bands.

Screen block or grille units have gained wide popularity as decorative and functional masonry. A few designs are shown in Fig. 1-24. The units available have a wide range of sizes, from 4 to 16 in. square, to meet nearly every need. Though often used mainly for their esthetic value, they also provide excellent balance between privacy and vision from within or without (Fig. 1-25). They diffuse strong sunlight, provide a wind break, and yet permit free flow of air. The decorative value of the units is enhanced by the effects that variations in light and shade produce on their patterns. The principal uses are in decorative building facades, ornamental room dividers and partitions, garden fences, and patio screens. Construction details are given in Chapter 8.

Although there is almost an endless variety of patterns for screen block, *the number available in any locality may be limited.* When a pattern is not available locally, it is often possible for the block plant to rent the mold from the block machine manufacturer or another block producer. The numerous screen block designs available, coupled with the possibility of using

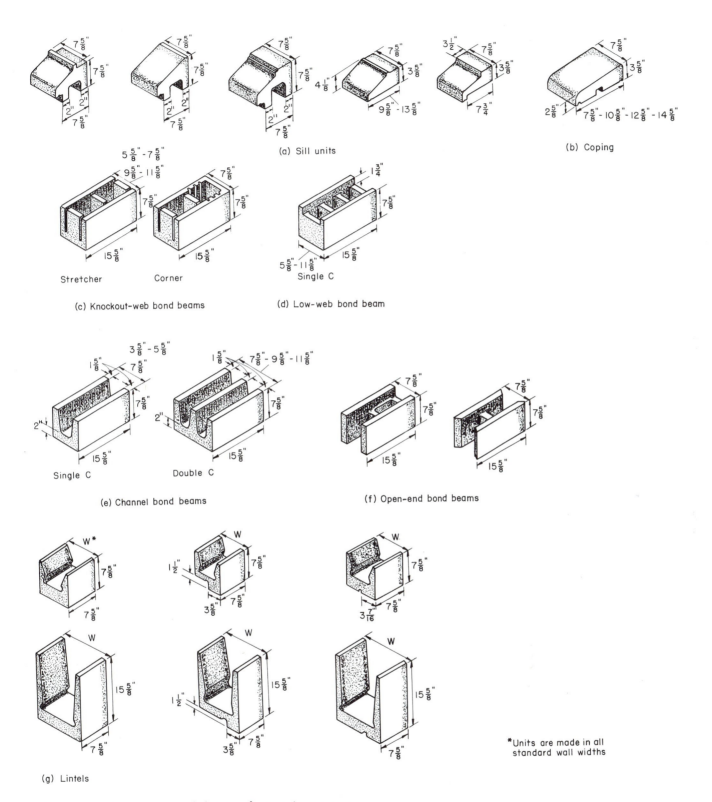

(a) Sill units

(b) Coping

Stretcher   Corner   Single C

(c) Knockout-web bond beams   (d) Low-web bond beam

Single C   Double C

(e) Channel bond beams   (f) Open-end bond beams

(g) Lintels

*Units are made in all standard wall widths

Fig. 1-18. Some sizes and shapes of concrete masonry units for sills, copings, bond beams, and lintels. *Consult local producers for specific available units.*

Fig. 1-19. Channel bond beam.

Fig. 1-20. Low-web bond beam.

several orientations of a particular unit in a wall, give the designer nearly unlimited opportunity for producing beautiful screen wall effects. Although a wide range of designs with screen block can be obtained, basically the designs include:

1. Units that are a complete pattern in themselves. When laid, the wall forms a panel of small individual repetitive patterns.
2. Units that form part of a pattern. The pattern may require two or sometimes four units to be completed. It is important to consider the esthetic effects of an incomplete pattern if the dimensions of the wall are not a multiple of the dimensions of the pattern.
3. Various types of units that can form an overall pattern in a wall with very interesting and varied effects. Severl different patterns are possible using only two types of these units.
4. Conventional solid or hollow block units, which can be used quite successfully in screen walls if laid with spaces between the units. Hollow block laid on their ends or sides also provide an interesting and attractive screen wall.

Architectural concrete masonry units, sometimes known as customized masonry or sculptured units, offer opportunities for almost unlimited architectural freedom. Some of the possibilities are indicated in Fig. 1-26. Patterns and profiles in the block can be achieved with vertical scoring, fluted or ribbed faces, molded angles or curves, projected or recessed faces (Fig. 1-27), or combinations of these surfaces. The designer may select virtually any shape that can be molded vertically within the bounds of the 18×26-in. metal pallet under the block machine. A few block machines have larger pallets. Usually the vertical height of the unit is limited to 7⅝ in., though some block manufacturers produce units 11⅝ in. high.

With architectural concrete masonry the selection of the unit profile desired may be closely related to the architectural design of the building. Also, the surface color and texture, whether or not aggregates will be exposed, the type of aggregates to be exposed, whether or not white or buff-colored cement or color pigments will be used, unit dimensions—all of these are selections that the architect may relate to design. Other effects also can be created by using the same block design in different pattern bonds. The play of light and shade on the profiled faces can be varied according to the projected or recessed position in which each block (of the same pattern) is laid. Furthermore, by use of sculptured units alone or in combination with plain units, a variety of geometric patterns in relief can be designed.

Other Sizes, Shapes, and Types

Block scoring (Fig. 1-28) refers to a process whereby any block can be given a new apparent face size by saw-cutting the face either horizontally or vertically, or by molding depressions into the face. Scored block are primarily used to achieve a ⅓ to ⅔ face height for ashlar patterns, but it is possible to allow the imagination to run rampant and achieve any number of patterns.

A patented slotted block (Fig. 1-29a) provides unusually high sound energy absorption. The slotted openings molded into the face of the units conduct sound into the cores, acting as damped resonators. Especially effective with sound in the middle and high frequencies, these block are very useful in gymnasiums, factories, bowling alleys, or other places where noise generation is high.

Special lightweight block are made of lightweight insulating concrete in a variety of sizes (Fig. 1-29b). The weight of an 8×12×24- or 12×8×24-in. unit is comparable to that of a normal-weight 8×8×16-in. concrete block.

A special H-shaped block (Fig. 1-29c) is made for reinforced grouted masonry. This unit can be easily

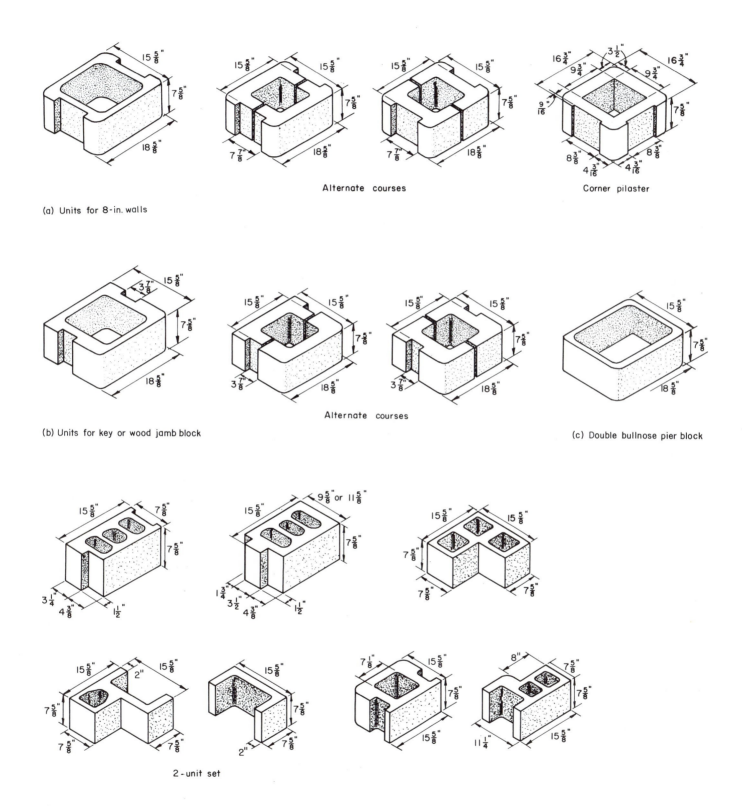

(a) Units for 8-in. walls

Alternate courses

Corner pilaster

(b) Units for key or wood jamb block

Alternate courses

(c) Double bullnose pier block

2-unit set

(d) Units for special conditions

Fig. 1-21. Pilaster units. *Consult local producers for specific available units.*

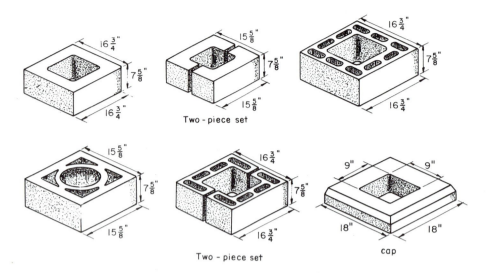

Fig. 1-22. Customized column and chimney units and chimney caps. *Consult local producers for specific available units.*

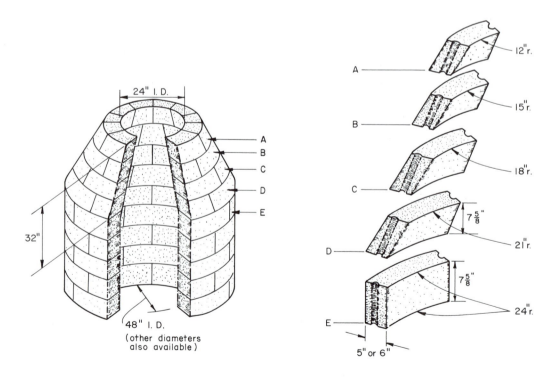

Fig. 1-23. Manhole, catch basin, and valve vault units. *Consult local producers for specific available units.*

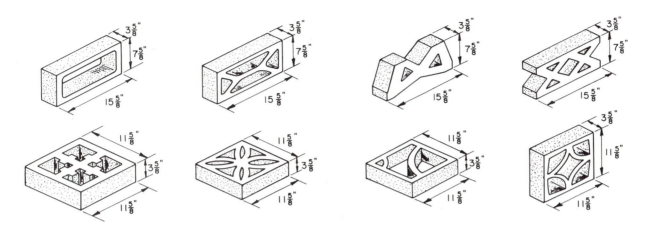

Fig. 1-24. Screen wall units. *Consult local producers for specific available units.*

Fig. 1-25. Screen walls give privacy at the bedrooms in these apartments.

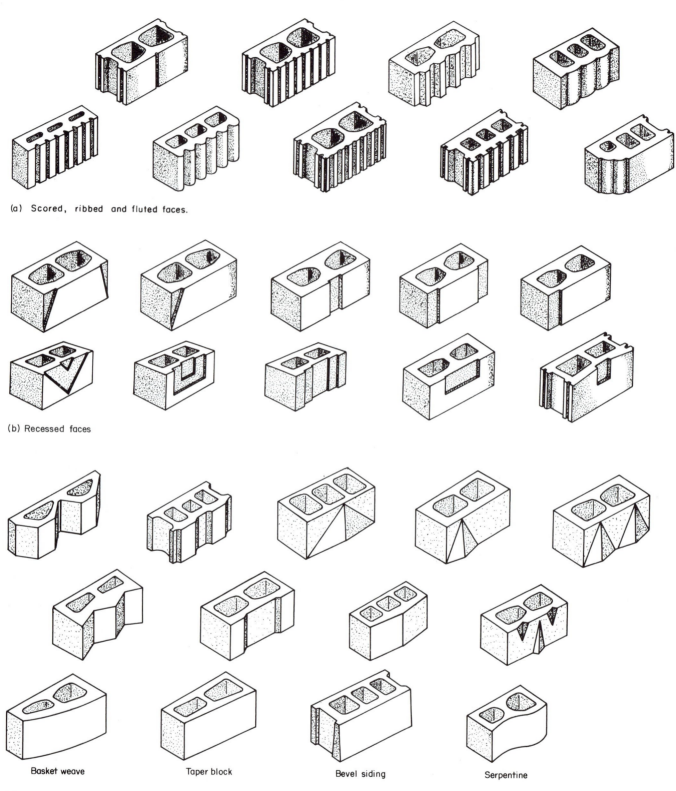

(a) Scored, ribbed and fluted faces.

(b) Recessed faces

Basket weave Taper block Bevel siding Serpentine

(c) Angular and curved faces

Fig. 1-26. Architectural concrete masonry units. *Consult local producers for specific available units.*

Fig. 1-27. Commercial building made with recessed-face block.

Fig. 1-28. Vertical scoring provides a dignified pattern for this office lobby.

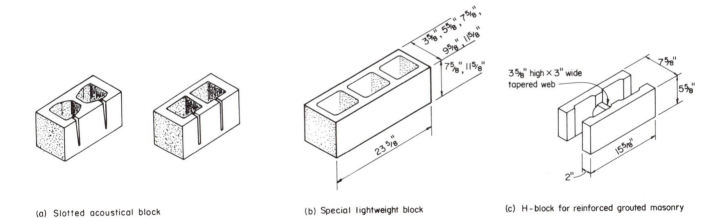

(a) Slotted acoustical block   (b) Special lightweight block   (c) H-block for reinforced grouted masonry

Fig. 1-29. Some special units. *Consult local producers for specific available units.*

placed between tall reinforcing bars.

Other specialized shapes for retaining walls, chimneys, foundations, and paving are discussed in Chapter 8.

Prefacing

Prefaced concrete masonry units offer opportunities for a wide range of colors, patterns, and textures. Prefaced units are sometimes supplied with scored or patterned surfaces—in a variety of colors. Colors may be even or the surface may be dappled. The surface may be prefaced on one or two sides, or sides and end, as required. They also may be produced in a variety of thicknesses, heights, and shapes, with cove bases and bullnose edges. A few examples of prefaced concrete masonry units are shown in Fig. 1-30.

The applicable specification for prefaced units is Standard Specification for Prefaced Concrete and Calcium Silicate Masonry Units, ASTM C744. The more common binders used to face masonry units are resin, resin and inert filler (fine sand), or portland cement and inert filler. Ceramic or porcelainized glazes and mineral glazes are also used, and these units are often called glazed concrete masonry.

There are numerous variations in the resin formulations and in the methods employed in producing resin-base facings. In some facings the resin-aggregate mixture is vibrated into flat pans, applied to faces of hardened block, and cured by application of heat. In others the facing is applied by spraying the face of the hardened block before heat processing.

With the cement-aggregate method, the facing mixture is prepared and vibrated into flat pans. These pans are then inserted into the block machine and the block is cast, with facing and block forming an integral bond. Generally, the heights and lengths of pan-applied facings are 1/8 in. greater than the heights and lengths of the regular modular concrete masonry units to which they are bonded. As a result, the exposed face of the mortar joints must be 1/4 in. thick to preserve modular dimensioning of the masonry wall.

Because the facings are resistant to water penetration, abrasion, and cleaning detergents, prefaced units are used in walls and partitions where decoration, cleanliness, and low maintenance are desirable. This is the case for wainscoting in school corridors and locker rooms, in bottling plants, in special areas of hospitals, and in food processing plants. On the other hand, architects have achieved striking effects by combining vividly colored glazed units with regular painted units or with the types of units represented in Fig. 1-26.

In some localities masonry units are available that are prefinished with colored glazes spray-applied in a manner which preserves the original surface texture and sound absorption properties. Thus, they are useful in constructing walls for auditoriums, churches, classrooms, lecture halls, and other areas where pleasing color effects, good acoustical properties, and minimal wall surface maintenance are sought.

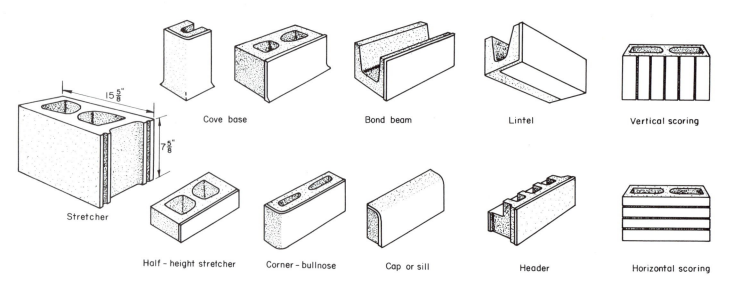

Fig. 1-30. Prefaced units. *Consult local producers for specific available units.*

Fig. 1-31. Glazed concrete masonry units are used at swimming pools where sanitation and a permanently attractive finish are needed.

Surface Texture

The surface texture of concrete masonry units may be greatly varied to satisfy esthetic requirements or to suit a desired physical requirement. Various degrees of smoothness can be acheived with any aggregate by changes in the aggregate grading, mix proportions, wetness of the mix, and the amount of compaction in molding.

Textures are classified somewhat loosely and with considerable overlapping as open, tight, fine, medium, and coarse. A texture regarded as fine in one locality may be considered medium in another.

An open texture is characterized by numerous closely spaced and relatively large voids between the aggregate particles. Conversely, a tight texture is one in which the spaces between aggregate particles are well filled with cement paste; it has few pores or voids of the size readily penetrated by water and sound.

Fine, medium, and coarse describe the relative smoothness or graining of the texture. A fine texture is not only smooth but made up of small, very closely spaced granular particles. A coarse texture is noticeably large-grained and rough, resulting from the presence of a large proportion of large-size aggregate particles in the surface. Usually, but not necessarily, it will contain substantial-size voids between aggregate particles. A medium texture is one intermediate between fine and coarse; it is an average texture. Examples of several of these textures are shown in Fig. 1-32.

If the concrete masonry surface is to serve as a base for stucco or plaster, a coarse texture is desirable for good bond. Coarse and medium textures provide sound absorption even when painted. The paint, however, must be applied in a manner that does not close all of the surface pores; spray painting is best. A fine texture is preferred for ease of painting.

In regular plant production the texture of concrete masonry units will be fairly uniform from day to day

Fig. 1-32. Examples of textures of concrete masonry. The top two units were made with lightweight aggregate, the bottom two with normal-weight aggregate.

Fig. 1-33. Ground-face units expose the natural color of the aggregate.

and shipment to shipment, but absolute uniformity is not attainable. The expected range in texture can best be determined from a sample of 10 or more units taken at random from the manufacturer's stockpile.

With a few exceptions manufacturers limit their regular production of concrete masonry units to a single texture for each aggregate type. Otherwise they could not operate economically because of the problems and expense connected with making, stockpiling, and merchandising several classes of units based on small differences in texture. Whether the texture adopted is fine, medium, or coarse will depend to a large extent upon local practice and preference. Some manufacturers will produce other textures on special order provided the order is large enough and the texture desired can be produced with the material and equipment available at a price acceptable to the customer.

A texturing process that may be selected is ground-facing (Fig. 1-33). Ground-face masonry units are produced from normal-weight or lightweight units by grinding off a 1/16- to 1/8-in. layer of concrete from one or both face shells. The process results in a smooth, open-textured surface that shows aggregate particles of varying color to good advantage. Variations in aggregate size, type, and color and the use of integral pigments offer many opportunities for adding interest to the surface in the form of color and texture. Ground-face units in natural or tinted colors are often used in constructing partition walls and corridor walls that are to be exposed without further finishing, except perhaps the application of wax or a colorless sealer. As an alternate to grinding, concrete masonry units may be sandblasted before or after they are placed.

Color

The natural color of concrete masonry varies from light to dark grey to tints of buff, red, or brown, depending upon the color of the aggregate, cement, and other mix ingredients used as well as the method of curing. Color uniformity of masonry units is not controlled unless by special customer order. There may be some variations in the color of any given day's production even though all conditions are apparently alike.

Units also are subject to temporary and permanent changes in color. Colored surfaces are more vivid and darker when wet than when dry. Units made with dark-colored aggregate will slowly become darker with age when subject to weathering because the surface film of cement paste erodes away, exposing the aggregate. While stockpiled at the plant or for long periods at the jobsite, units may undergo slight color changes due to dust and soot lodging in the surface pores. Despite the numerous factors that may affect color, units of a specific type of aggregate and method of curing are generally uniform in color within acceptable limits.

This is particularly true when the units for the project or building are from the same manufacturer.

Much concrete masonry construction is painted for architectural or service exposure reasons (see Chapter 7) and so uniform color of the block in these cases is not an important factor. On the other hand, highly pleasing effects have been achieved in building interiors with delicate differences in shades of color of unpainted units (Fig. 1-34). Obviously, such units should be free of stains and other blemishes.

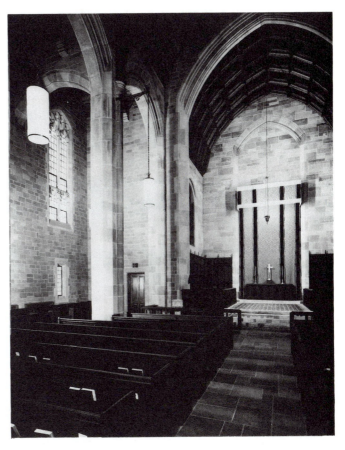

Fig. 1-34. Quiteness and beauty are achieved in this church with walls of concrete masonry.

In recent years a noticeable trend is developing toward the use of colored mortar joints for laying concrete masonry. Many architects prefer mortar joints that are identical to or harmonize with the shade of color of the masonry units involved. Colored mortar is discussed in Chapter 2.

The use of locally available natural sands, cements, and coarse aggregates to produce the desired color is recommended where possible. This will result in a more easily duplicated color in the event of future additions to a structure.

Integral coloration of concrete masonry units is possible through the use of mineral oxide pigments mixed with the concrete before molding. An ever-increasing variety of colored concrete masonry products are being offered. This trend started with pigmented standard building block and has since spread to such products as concrete brick, split block, slump block, paving and patio block, screen block, and architectural concrete masonry units.

Standard colors for integrally colored concrete masonry units are tan, buff, red, brown, pink, yellow, and black or grey. Green can be produced and is quite permanent but expensive, except in light shades. Blue is expensive and not uniform or permanent.

Pigmented concrete masonry units should not be stored in the open. If they are stacked on their sides, greater drying is possible and any soluble lime can combine with carbon dioxide from the air to reduce efforescence. Nonuniformity of coloring can be counteracted by distributing concrete masonry units to random locations on the job, thus intermingling the different shades of the units.

Chapter 2

MORTAR AND GROUT

Mortar for concrete masonry (Fig. 2-1) not only joins masonry units into an integral structure with predictable performance properties, but also: (1) effects a tight seal between units against the entry of air and moisture; (2) bonds with joint reinforcement, metal ties, and anchor bolts, if any, so that they perform integrally with the masonry; (3) provides an architectural quality to exposed masonry structures through color contrasts or shadow lines from various joint-tooling procedures (Fig. 2-2); and (4) compensates for size variations in the units by providing a bed to accommodate dimensional tolerances of units.

Grout is an essential element of reinforced concrete masonry. In reinforced load-bearing masonry wall construction, grout is usually placed only in those wall spaces containing steel reinforcement. The grout bonds the masonry units and steel so that they act together to resist imposed loads. In some reinforced load-bearing masonry walls, all cores—with and without reinforcement—are grouted to further increase the wall resistance to loads. Grout is sometimes used in nonreinforced load-bearing masonry wall construction to give added strength. This is accomplished by filling a portion or all of the cores.

Mortar is not grout. Grout and mortar are used differently, have different characteristics, and are handled differently. Thus, the two are not interchangeable.

Mortar

Masonry mortar is composed of one or more cementitious materials, clean well-graded masonry sand, and sufficient water to produce a plastic, workable mixture. By modern specifications the ratios range, by volume, from 1 part of cementitious material to 2¼ to 3½ parts of damp, loose mortar sand.

Probably the most important quality of a masonry mortar is workability because of its influence on other important mortar properties in both the plastic and

Fig. 2-1. Mortar is an integral part of concrete masonry.

Fig. 2-2. Split block in an apartment building. Mortar and the joints have a subtle influence on the architecture.

hardened states. Since the mortar is an integral part of a concrete masonry structure and some characteristics of mortar materially affect the quality of workmanship obtained, the mortar should be designed and specified with the same care as the masonry unit itself.

Desirable Properties of Fresh, Plastic Mortar

Good mortar is necessary for good workmanship and proper structural performance of concrete masonry. Since mortar must bond masonry units into strong, durable, weathertight structures, it must have many desirable properties and the materials must comply with specifications. Desirable properties of mortar while plastic include workability, water retentivity, and a consistent rate of hardening.

Workability

This property of plastic mortar is difficult to define because it is a combination of a number of independent, interrelated properties. The interrelated mortar properties considered as having the greatest influence on workability are: consistency, water retentivity, setting time, weight, adhesion, and cohesion.

The experienced mason judges the workability of mortar by the way it adheres to or slides from his trowel. Mortar of good workability should spread easily on the concrete masonry unit, cling to vertical surfaces, extrude readily from joints without dropping or smearing, and permit easy positioning of the unit without subsequent shifting due to its weight or the weight of successive courses (Fig. 2-3). Mortar consistency should change with weather to help in laying the units. A good workable mix should be softer in summer than in winter to compensate for water loss.

Fig. 2-3. Mortar of proper workability is soft but with good body; it spreads readily and extrudes without smearing or dropping away.

Water Retentivity

This is the property of mortar that resists rapid loss of mixing water (prevents loss of plasticity) to the air on a dry day or to an absorptive masonry unit. Rapid loss of water causes the mortar to stiffen quickly, thereby making it practically impossible to obtain good bond and weathertight joints.*

Water retention is an important property and related to workability. A mortar that has good water retentivity remains soft and plastic long enough for the masonry units to be carefully aligned, leveled, plumbed, and adjusted to proper line without danger of breaking the intimate contact or bond between the mortar and the units. When low-absorption units such as split block

*See Ref. 26.

are in contact with a mortar having too much water retentivity, they may float. Consequently, the water retention of a mortar should be within tolerable limits.

Entrained air, extremely fine aggregate or cementitious materials, or water adds workability or plasticity to the mortar and increases its water retentivity.

Consistent Rate of Hardening

Mortar hardens when cement reacts chemically with water in a process called hydration. The rate of hardening of mortar is the speed at which it develops resistance to an applied load. Too rapid hardening may interfere with the use of the mortar by the mason. Very slow hardening may impede the progress of the work since the mortar will flow from the completed masonry. During winter construction, slow hardening may also subject mortar to early damage from frost action. A well-defined, consistent rate of hardening assists the mason in laying the masonry units and in tooling the joints at the same degree of hardness. Uniform joint color of masonry reflects proper hardening and consistent tooling times.

Hardening is sometimes confused with a stiffening caused by rapid loss of water, as in the case of low-water-retention mortars with high absorptive units. Also, mortar tends to stiffen more rapidly than usual during hot and dry weather. In this case the mason may find it advisable to lay shorter mortar beds and fewer units in advance of tooling.

Desirable Properties of Hardened Mortar

Durability

The durability of masonry mortar is its ability to endure the exposure conditions. Although aggressive environments and the use of unsound materials may contribute to the deterioration of mortar joints, the major destruction is from water entering the concrete masonry and freezing.

In general, damage to mortar joints and to mortar bond by frost action has not been a problem in concrete masonry wall construction above grade. In order for frost damage to occur, the hardened mortar must first be water-saturated or nearly so. After being placed, mortar becomes less than saturated due to the absorption of some of the mixing water by the units. The saturated condition does not readily return except under special conditions, such as: (1) the masonry is constantly in contact with saturated soils; (2) downspouts leak; (3) there are heavy rains; or (4) horizontal ledges are formed. Under these conditions the masonry units and mortar may become saturated and undergo freeze-thaw deterioration.

High-compressive-strength mortars usually have good durability. Because air-entrained mortar will

withstand hundreds of freeze-thaw cycles, its use provides good insurance against localized freeze-thaw damage. Masonry cement mortar has a higher air content than portland cement and lime mortar and therefore has better freeze-thaw resistance than non-air-entrained portland cement and lime mortar. Mortar joints deteriorated due to freezing and thawing present a maintenance problem generally requiring tuckpointing.

Sulfate resistance is usually not a concern for masonry above ground; however, in some parts of the country, masonry can be exposed to sulfate from soil, ground water, or industrial processes. Sulfate-resistant masonry materials should be used when in contact with soils containing more than 0.1% water-soluble sulfate (SO_4) or water solutions containing more than 150 ppm of sulfate. Without the use of sulfate-resistant masonry units and mortar or use of a protective treatment, sulfates attack and deteriorate masonry. Masonry cements, sulfate-resistant portland cements (Type II or V), or sulfate-resistant blended cements should be used in mortar exposed to sulfates. One study demonstrated that masonry cement is significantly more sulfate resistant than a Type II cement and lime mortar when tested in accordance with ASTM C1012 (at 13 weeks, portland cement-lime mortars exhibited expansions of 0.16% to 0.37% compared to 0.03% to 0.12% for masonry cement mortars). See Reference 71 for more information.

Compressive Strength

The principal factors affecting the compressive strength of concrete masonry structures are the compressive strength of the masonry unit, the proportions of ingredients within the mortar, the design of the structure, the workmanship, and the degree of curing. Although the compressive strength of concrete masonry may be increased with a stronger mortar, the increase is not proportional to the compressive strength of the mortar.

Tests have shown that concrete masonry wall compressive strengths increase only about 10% when mortar cube compressive strengths (Fig. 2-4) increase 130%. Composite wall compressive strengths increase 25% when mortar cube compressive strengths increase 160%.

Compressive strength of mortar is largely dependent on the type and quantity of cementitious material used in preparing the mortar. It increases with an increase in cement content and decreases with an increase in air entrainment, lime content, or water content.

Portland cement in concrete masonry mortar requires a period in the presence of moisture to develop its full strength potential. In order to obtain optimum curing conditions, the mortar mixture should have the maximum amount of mixing water possible with acceptable workability, considering maximum water retention; that is, lean, oversanded mixtures should be avoided. Freshly laid masonry should be protected from the sun and drying winds. With severe drying

Fig. 2-4. Cube compressive strength test.

conditions it may be necessary either to fog the exposed mortar joints with a fine water spray daily for about 4 days or to cover the masonry structure with a polyethylene plastic sheet, or to do both.

Bond

The general term "bond" refers to a specific property that can be subdivided into: (1) the extent of bond, or the degree of contact of the mortar with the concrete masonry units; and (2) the bond strength, or the force required to separate the units. A chemical and a mechanical bond exist in each category.

Good extent of bond (complete and intimate contact) is important to watertightness and bond strength. Bond strength is defined and tested as tensile or flexural bond strength. In determining direct tensile bond strength, specimens representing unit and mortar are pulled apart (Fig. 2-5). Test methods for measuring flexural (more properly termed flexural tensile) bond strength place a more complex load on the mortar-to-unit interface, but can be applied to full-sized specimens. For example, ASTM Method C952 (Fig. 2-6) utilizes a bond wrench apparatus and loading configuration to induce failure of prisms constructed from full-sized masonry units.

Bond is a function of mortar, unit, and workmanship. While bond strength is an important property of masonry, current methods of test for determining bond strength are considered impractical as a basis for material specifications or quality control at the jobsite due to the high variability of results inherent in the testing methods.

Poor bond at the mortar-to-unit interface may lead to moisture penetration through the unbonded areas. Good extent of bond is obtained with a workable and water-retentive mortar, good workmanship, full joints,

Fig. 2-5. Tensile bond test.

Fig. 2-6. ASTM C952 bond wrench test.

and concrete masonry units having a medium initial rate of absorption (suction).

Mortar must develop sufficient bond to withstand the tensile and flexural forces brought about by structural, earth, and wind loads; shrinkage of concrete masonry units or mortar; and temperature changes. For laboratory fabricated specimens, the bond strength of mortar to masonry units can be determined in accordance with ASTM C952. The flexural bond strength of mortar and masonry unit combinations, using either laboratory or field fabricated specimens or specimens cut from existing masonry, can be determined from ASTM C1072.

Many variables affect bond, including: (1) mortar ingredients, such as type and amount of cementitious materials, water retained, and air content; (2) characteristics of the masonry units, such as surface texture, suction, and moisture content; (3) workmanship, such as pressure applied to mortar bed during placing; and (4) curing conditions, such as temperature, relative humidity, and wind. ASTM C952 and C1072 can be used to analyze the effect on bond strength of job materials and conditions, however, these results should not be interpreted as the bond strength of the masonry structure. The general effects of some of these variables on bond are discussed below.

All other factors bring equal, mortar bond strength is related to mortar composition, especially the cement content. Bond strength of mortar increases as cement content increases.

There is a direct relationship between mortar flow (water content) and tensile bond strength. For all mortars, bond strength increases as water content increases. The optimum bond strength is obtained while using a mortar with the highest water content compatible with workability, even though mortar compressive strength decreases.

Workmanship is paramount in affecting bond strength. The time lapse between the spreading of mortar and the placing of the concrete masonry units should be kept to a minimum because the water content of the mortar will be reduced through absorption by the masonry unit on which it is first placed. If too much time elapses before another unit is placed upon it, the bond between the mortar and the unit being placed will be reduced. The mason should not be permitted to realign, tap, or in any way move units after initial placement, leveling, and alignment. Movement breaks the bond between unit and mortar, after which the mortar will not readhere well to the masonry units.

Volume Change

A popular misconception is that mortar shrinkage can be extensive and cause leaky structures. Actually, the maximum shrinkage across a mortar joint with properly proportioned mortar is usually miniscule and therefore not troublesome. This is even truer with the weaker mortars. They have greater creep, that is, extensibility, and so are better able to accommodate shrinkage.

As available water in mortar is absorbed by the masonry units and lost through evaporation, some drying shrinkage occurs. Though generally not a problem in masonry construction, exteme drying shrinkage can result in development of cracks in the mortar. Since drying shrinkage is related to the amount of water lost by the mortar, factors that increase water content of a mortar tend to increase its drying shrinkage. For example, air-entrained mortars tend to have a lower water demand than non-air-entrained mortars at an equivalent flow and thus exhibit less drying shrinkage. However, this principle should not be misinterpreted to mean that water content of mortar should be arbitrarily reduced. As previously noted, workability and bond are directly related to the flow of the mortar and should be given priority in determining the water content of field mixed mortar.

On projects where it is desirable to minimize drying shrinkage, masonry cement mortar should be considered. The shrinkage of mortar can be tested in accordance with ASTM C1148. In a study using this test, masonry cement mortar had half the shrinkage of cement-lime mortar (0.07% at 25 days for masonry cement mortar versus 0.12% to 0.14% for cement-lime mortar). See Reference 71 for more information.

Expansion in mortars due to unsound ingredients can cause serious disintegration of masonry. Soundness of a cementitious material is measured by the autoclave expansion test (ASTM C91). This test produces reactions in unsound ingredients (particularly free lime and periclase) and simulates a long period of exposure in a wall. Any changes in length of the test specimen are measured to ascertain that there will be no serious expansion of the hardened mortar in the wall. While a method for measuring soundness of hydrated lime has been developed, correlation of results to field performance has not yet been established. Thus soundness of this material is generally assured by limiting the unhydrated oxide content of the hydrated lime to a maximum of 8%.

Appearance

Uniformity of color and shade of the mortar joints greatly affects the overall appearance of a concrete masonry structure (Figs. 2-2 and 2-7). Atmospheric conditions, admixtures, and moisture content of the masonry units are some of the factors affecting the color and shade of mortar joints. Others are uniformity of proportions in the mortar mix, water content, and time of tooling the mortar joints.

Careful measurement of mortar materials and thorough mixing are important to maintain uniformity from batch to batch and from day to day. Control of this uniformity becomes more difficult with the num-

Fig. 2-7. Ribbed split block wall.

ber of ingredients to be combined at the mixer. Pigments, if used, will provide more uniform color if premixed with a stock of cement sufficient for the needs of the whole project. In many areas, colored masonry cements are available; they provide better control over color uniformity.

Tooling of mortar joints at like degrees of setting is important in ensuring a uniform mortar shade in the finished structure. If the joint is tooled when the mortar is relatively hard, a darker shade results than if the joints are tooled when the mortar is relatively soft. Some masons consider mortar joints ready for tooling after the mortar has stiffened but is still thumbprint-hard, with the water sheen gone.

White cement mortar should never be tooled with metal tools because the metal will darken the joint. A glass or plastic joint tool should be used.*

Specifications and Types

Mortars are selected and prepared in accordance with either the property or the performance specifications of ASTM C270, Specification for Mortar for Unit Masonry. Conformance with this specification ensures coupling individual materials conforming to product specifications with industry recommendations. It is the

practice in the masonry industry to dissuade the use of higher strength mortars than specified and to use the maximum quantity of water and sand to produce the desired mortar.

Mortar has traditionally been classified as Types M, S, N, O, and K (every other letter from the words "MASON WORK"). Type K is not used in new construction, but may be used in certain tuckpointing applications (see "Tuckpointing," Chapter 9).

The current specifications for mortars for unit masonry are shown in Tables 2-1 and 2-2. Mortar types are identified by proportion or property specification, but not both.

The proportion specification identifies mortar type through various combinations of portland cement with masonry cement, masonry cement singly, and combinations of portland cement and lime. The proportion specification governs when ASTM C270 is referred to without noting which specification—proportion or property specification—shall be used.

Mortar type classification under the property specification is dependent primarily on the compressive strength of 2-in. cubes using laboratory tests (Fig. 2-4) per ASTM C270. These laboratory test cubes are pre-

---

*For a further discussion of tooling as well as the types of mortar joints, see "Tooling Mortar Joints," Chapter 6.

Table 2-1. Proportion Specifications for Mortar*

Mortar type	Parts by volume					
	Portland cement or blended cement	Masonry cement type			Hydrated lime or lime putty	Aggregate
		M	S	N		
M	1	—	—	1	—	4½ to 6
	—	1	—	—	—	2¼ to 3
	1	—	—	—	¼	2¹³/₁₆ to 3¾
S	½	—	—	1	—	3⅜ to 4½
	—	—	1	—	—	2¼ to 3
	1	—	—	—	Over ¼ to ½	**
N	—	—	—	1	—	2¼ to 3
	1	—	—	—	Over ½ to 1¼	**
O	—	—	—	1	—	2¼ to 3
	1	—	—	—	Over 1¼ to 2½	**

*Adapted from ASTM C270.
**The total aggregate shall be equal to not less than 2¼ and not more than 3 times the sum of the volumes of the cement and lime used.

Notes: 1. Under ASTM C270, Standard Specification for Mortar for Unit Masonry, aggregate is measured in a damp, loose condition and 1 cu.ft. of masonry sand by damp, loose volume is considered equal to 80 lb. of dry sand.
2. Mortar should not contain more than one air-entraining material.

Table 2-2. Property Specifications for Laboratory-Prepared Mortar*

Mortar type	Minimum compressive strength, psi	Minimum water retention, %	Maximum air content, %**
	At 28 days		
M	2500	75	12†
S	1800	75	12†
N	750	75	14†
O	350	75	14†

*Adapted from ASTM C270.
**Cement-lime mortar only (except where noted).
†When structural reinforcement is incorporated in cement-lime or masonry cement mortar, the maximum air content shall be 12% or 18%, respectively.

Note: The total aggregate shall be equal to not less than 2¼ and not more than 3½ times the sum of the volumes of the cement and lime used.

pared with less water than will be used on the job. Similar cube tests are not intended to be made on the job. Instead, mortar may be tested in the field according to Standard Method for Pre-Construction and Construction Evaluation of Mortars for Plain and Reinforced Unit Masonry (ASTM C780).*

The ratio of cementitious material to aggregate in the mixture under the property specification may be less than under the proportion specification. This is to encourage preconstruction mortar testing; an econom-ic reward is possible if less cement is required in a mix to meet the strength requirement of the property specification. The testing portion of this specification is limited to preconstruction evaluation of the mortars.

In both the proportion and property specifications, the amount of water to be used on the job is the maximum that will produce a workable consistency during construction. This is unlike conventional concrete practice where the water-cement ratio must be carefully controlled.

Another requirement of property specifications is the water retention limit. In the laboratory it is measured using a "flow-after-suction" test (described in ASTM C91, Standard Specification for Masonry Cement), which simulates the action of absorptive masonry units on the plastic mortar. In performance of a flow test before and after absorptive suction, a truncated cone of mortar is subjected to twenty-five ½ in. drops of a laboratory flow table (Fig. 2-8). The diameter of the disturbed sample is compared to the original diameter of the conical mortar sample. Allowable initial flow tests range from 105 to 115% and the flow-after-suction test must exceed 75%. These values are specified for laboratory test purposes while flow values of 130 to 150% are common for mortar in actual construction.

An interplay of proportion and property specifications is not intended or recognized by the specifications.

_____
*See "Testing Project Mortar," Chapter 6.

Fig. 2-8. Flow test.

Once the design loads, type of structure, and masonry unit have been determined, the mortar type can be selected. No one mortar type will produce a mortar that rates highest in all desirable mortar properties. Adjustments in the mix to improve one property often are made at the expense of others. For this reason, the properties of each mortar type should be evaluated and the mortar type chosen that will best satisfy the end-use requirements.

For plain (nonreinforced) masonry, mortar type is selected by the architect-engineer team on the basis of Table 2-3. For reinforced masonry, the job specification should state the mortar requirements. Freedom of selection of individual mixtures is favored since workability of mixtures and availability of cementitious materials vary with geographical areas.

From the preceding paragraphs it can be seen that a choice of mortar type still exists. Where an engineering analysis is requried, the type of mortar selected will, in conjunction with the compressive strength of the concrete masonry units, determine the allowable stresses for the wall. It is not always necessary to use a Type M mortar for high strength because many building codes rate Type S in the wall as giving equally high strength. Moreover, Type S or N mortar has more workability, water retention, and extensibility.

The choice of using masonry cement or a portland cement and lime combination is largely a matter of economics and convenience. Either will produce mortar with acceptable properties as long as the specifications are met. Masonry cements provide all cementitious materials required for a masonry mortar in one bag. The quality and appearance of mortars made from masonry cement are consistent because the masonry cement materials are mixed and ground together before being packaged. Consequently, masonry cement mortars are less subject to variations from batch to batch than mortars produced from combining ingredients on the job.

Colors

Pleasing architectural effects with color contrast or harmony between masonry units and joints are obtained through the use of white or colored mortars.

Table 2-3. Guide to the Selection of Mortar Type*

Location	Building segment	Mortar type	
		Recommended	Alternative
Exterior, above grade	Load-bearing walls Non-load-bearing walls Parapet walls	N O** N	S or M N or S S
Exterior, at or below grade	Foundation walls, retaining walls, manholes, sewers, pavements, walks, and patios	S†	M or N†
Interior	Load-bearing walls Non-load-bearing partitions	N O	S or M N

*Adapted from ASTM C270. This table does not provide for specialized mortar uses, such as chimney, reinforced masonry, and acid-resistant mortars.

**Type O mortar is recommended for use where the masonry is unlikely to be frozen when saturated or unlikely to be subjected to high winds or other significant lateral loads. Type N or S mortar should be used in other cases.

†Masonry exposed to weather in a nominally horizontal surface is extremely vulnerable to weathering. Mortar for such masonry should be selected with due caution.

Note: For tuckpointing mortar, see ''Tuckpointing,'' Chapter 9.

White mortar is made with white masonry cement, or with white portland cement and lime, and white sand. For colored mortars, the use of white masonry cement or white portland cement instead of the normal grey cements not only produces cleaner, brighter colors but is essential for making pastel colors such as buff, cream, ivory, pink, and rose.

Integrally colored mortar may be obtained through use of color pigments, colored masonry cements, or colored sand. Brilliant or intense colors are not generally attainable in masonry mortars. The color of the mortar joints will depend not only on the color pigment, but also on the cementitious materials, aggregate, and water-cement ratio.

The property specification is often used instead of the proportion specification for colored mortar in order to more easily achieve the desired color by adjusting the mortar ingredients. The property specification also assures that the pigment is not adversely affecting mortar properties.

Pigments must be thoroughly dispersed throughout the mix. To determine if mixing is adequate, some of the mix is flattened under a trowel. If streaks of color are present, additional mixing is required. For best results, the pigment should be premixed with the cement in large controlled quantities. Colored masonry cement is available in many areas as Types M, S, and N.

As a rule, color pigments should be of mineral oxide composition and contain no dispersants that will slow or stop the portland cement hydration. Mineral oxides should not exceed 5% or 10% by weight of cement for masonry cement or cement-lime mortars, respectively. Iron, manganese, chromium, and cobalt oxides also have been used successfully. Zinc and lead oxides should be avoided because they may react with the cement. Carbon black may be used as a coloring agent to obtain dark grey or almost black mortar, but lampblack should not be used. Carbon black should be limited to 1% or 2% by weight of cement for masonry cement or cement-lime mortars, respectively; durability of this mortar may be deficient. In addition, the color of mortar using carbon black pigment rapidly fades with exposure to weathering.

It is recommended to use only those pigments that have been found acceptable by testing and experience. The following is a guide to the selection of coloring materials:

Red, yellow, brown, black or grey ....Iron oxide
Green .... Chromium oxide
Blue ....Cobalt oxide

Only the minimum quantity of pigment that will produce the desired shade should be used. An excess of pigment, more than 10% of the portland cement by weight, may be detrimental to the strength and durability of the mortar. The quantity of water used in mixing colored mortar should be accurately controlled.

The more water, the lighter the color. Retempering or the addition of water while using colored mortar should be avoided. Thus, any mortar not used while plastic and workable should be discarded.

Variations in the color of the materials are such as to make a color formula only approximate. Best results are obtained by experiment. Test panels should be made using the same materials and proportions as intended for use in the actual work, and the panels stored under conditions similar to those at the jobsite for about 5 days. Panels will have a darker shade when wet than when dry.

Color variation can be minimized by careful planning before the colored mortar is prepared at the project during construction. Some considerations include the use of preweighed bags of coloring material; adding pigments to the batch in a consistent manner; using whole bags of ingredients instead of partial bags if possible; using colored masonry cement; using ready-mixed mortar; or using onsite silo-type batching equipment.

Fading of colored mortar joints may be caused by efflorescence, the formation of a white film on the surface (Fig. 2-9). Efflorescence is more visable on a colored surface. The white deposits are caused by soluble salts that have emerged from below the surface, or by calcium hydroxide liberated during the setting of the cement and then combining with atmospheric carbon dioxide forming carbonate compounds. Good color pigments do not effloresce or contribute to efflores-

Fig. 2-9. A severe case of efflorescence.

cence. Efflorescence may be removed with a light sand-blasting, a stiff-bristle brush, or with mild cleaning techniques that are compatible with the mortar and masonry unit used. Cleaning of a selected trial area on site is recommended to evaluate which method and technique is best suited for removal of the efflorescence. Often dry brushing or cleaning with water and a brush are sufficient. See "Efflorescence," Chapter 9, for more information.

Components

Cementitious Materials

Foremost among the factors that contribute to good mortar is the quality of the mortar ingredients. The following material specifications of ASTM are applicable:

- Masonry cement—ASTM C91 (Types M, S, or N)
- Portland cement—ASTM C150 (Types I, IA, II, IIA, III, or IIIA)
- Blended hydraulic cement—ASTM C595[Types IS, IS-A, IP, IP-A, I(PM), or I(PM)-A]*
- Hydrated lime for masonry purposes—ASTM C207 (Types S, SA, N, or NA)**
- Quicklime for structural uses (for lime putty)—ASTM C5

Masonry Sand

The quantity of sand required to make 1 cu.ft. of mortar may be as much as 0.99 cu.ft.; hence, the sand has considerable influence on the mortar properties. Masonry sand for mortar should comply with the requirements of ASTM C144 (Standard Specification for Aggregate for Masonry Mortar). This specification includes both natural and manufactured sands. Sand should be clean, well-graded, and meet the gradation requirements listed in Table 2-4.

Table 2-4. Gradation for Mortar Sand

| Sieve size No. | Gradation specified, percent passing | |
| | ASTM C144* | |
	Natural sand	Manufactured sand
4	100	100
8	95 to 100	95 to 100
16	70 to 100	70 to 100
30	40 to 75	40 to 75
50	10 to 35	20 to 40
100	2 to 15	10 to 25
200	—	0 to 10

*Additional requirements: Not more than 50% shall be retained between any two sieve sizes nor more than 25% between No. 50 and No. 100 sieve sizes. In those cases where an aggregate fails to meet the gradation limit specified, it may be used if the masonry mortar will comply with the property specification of ASTM C270 (Table 2-2).

Sands with less than 5 to 15% passing the Nos. 50 and 100 sieves generally produce harsh or coarse mortars possessing poor workability; they also result in mortar joints with low resistance to moisture penetration. On the other hand, sands finer than those permitted by the above specification yield mortars with excellent workability, but they are weak and porous.

When mortar joints are specified to be less than the conventional ⅜-in. thickness, 100% of the sand should pass the No. 8 sieve and 95% the No. 16 sieve. For joints specified to be thicker than ⅜-in., the mortar sand selected should have a fineness modulus† approaching 2.5 or a gradation within the limits of concrete sands (fine aggregate) shown in ASTM C33, Standard Specification for Concrete Aggregates.

Water

Water intended for use in mixing mortar should be clean and free of deleterious amounts of acids, alkalies, and organic materials. Some potable waters contain appreciable amounts of soluble salts such as sodium and potassium sulfate. These salts can contribute to efflorescence later. Also, a water containing sugar would retard the set. Thus, the water should be fit to drink but investigated if it contains alkalies, sulfates, or sugars.

Admixtures

Although water-reducers, accelerators, retarders, and other admixtures are used in concrete construction, their use in masonry mortar may produce adverse effects on the normal chemical reaction between cement and water, especially during the early periods after mixing and when the water is available for hydrating the portland cement.

Hardened properties are also affected. To avoid metal corrosion, admixtures containing chloride (such as calcium choloride) or admixtures containing other corrosive substances must not be used. Set-controlling admixtures have been used reliably to maintain ready-mixed mortar (ASTM C1142) in a plastic, workable state for 24 to 72 hours. Regular retarders, as used in concrete, are particularly undesirable because they reduce strength development and increase the potential toward efflorescence. Air-entraining admixtures and accelerators are discussed later (Chapter 5) as admixtures for cold-weather masonry construction. Also, see "Modified Mortars" discussed later in this chapter.

---

*Slag cement Types S or SA can also be used but only in property specifications.

**Types N and NA lime may be used only if tests or performance records show that these limes are not detrimental to the soundness of mortar.

†Fineness modulus equals the sum of the cumulative percentages *retained* on the standard sieves, divided by 100. The higher the fineness modulus, the coarser the sand.

Whenever admixtures are considered for use in masonry, it is recommended that the admixture be laboratory-tested in the construction mortars at the temperature extremes requiring their use.

Material Storage

All cementitious materials, aggregates, and admixtures should be stored in such a manner as to prevent wetting, deterioration, or intrusion of foreign material. Brands of cementitious materials, admixtures, and the source of supply of sand should remain the same throughout the entire job.

Measuring Mortar Materials

Measurement of masonry mortar ingredients should be completed in a manner that will ensure the uniformity of mix proportions, yields, workability, and mortar color from batch to batch.

Aggregate proportions are generally expressed in terms of loose volume, but experience has shown that the amount of sand can vary due to moisture bulking. Fig. 2-10 shows how loose sand with varying amounts of surface moisture occupies differing volumes. Fig. 2-11 has the same data in another form for the fine and coarse sands, and shows the density of the sand. Loose, damp sand may consist of from 76 to 105 pcf of sand itself, plus the weight of the water. ASTM C270 states that a cubic foot of loose, damp sand contains 80 lb. of dry sand.

Ordinary sands will absorb water amounting to 0.4 to 2.3% of the weight of the sand. In the field, damp sands usually have 4 to 8% moisture, and so most of the water is on the surface of the sand.

Aside from the sand, other mortar ingredients are often sold in bags labeled only by weight. Since mortar is proportioned by volume, it is necessary to know:

	Unit weight, pcf
Portland cement	94
Blended cement	85 to 94*
Masonry cement	70 to 90*
Hydrated lime (dry)	40
Hydrated lime (putty)	80

The usual practice of measuring sand by the shovel can result in oversanding or undersanding or inconsistent batching of mortar. For more positive control it is recommended that accurate volume measurement of the sand be maintained at the jobsite by use of cubic foot measuring boxes. A hinged attachment of the measuring box to the mortar mixer can be constructed to facilitate one-man operation. As an alternative, calibrated five-gallon buckets may be used to measure the sand. For example, 1 cu.ft. is equal to 7.48 gallons, thus 3 cu.ft. of sand is 22.44 gallons or 4½ five-gallon buckets of sand. The following method is also suggest-

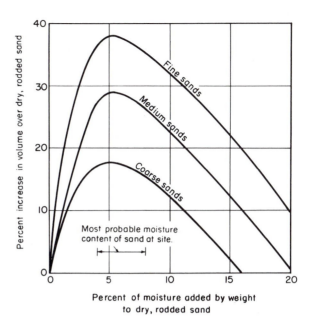

Fig. 2-10. Volume of loose, damp sand. PCA Major Series 172.

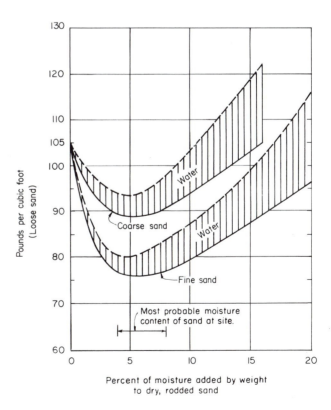

Fig. 2-11. Weight of loose, damp sand. PCA Major Series 172.

*See weight on bag.

ed: Construct one or two wooden boxes 12 in. square and 6 in. deep and use them to measure the sand required in a batch. Add the cement or lime by the bag. Then add water, measuring by the pail. When the desired consistency of mix is determined, mark the level of the mortar in the mixing drum. Use that as the mark for later batches when sand will be added by the shovelful. Repeat the measuring process halfway through the day or whenever the inspector requests it.

When sand is used in the dry condition, it will absorb some of the mix water during mixing and during the early period after mixing. If the mixer operator is not attentive to this natural occurrence, retempering of the mortar will be necessary to maintain workability and attain good bond.

Special techniques that overcome mix adjustment and jobsite variability problems include dry batching, packaged dry mortar, ready-mixed mortar, and silo-mixed mortar. These techniques are presented later in this chapter.

Mixing of Mortar

To obtain good workability and other desirable properties of plastic masonry mortar, the ingredients must be thoroughly mixed.

Mixing by Machine

Except possibly on very small jobs, mortar should be machine-mixed (Fig. 2-12). A typical mortar mixer has a capacity of 4 to 7 cu.ft. Conventional mortar mixers are of rotating spiral or paddle-blade design with tilting drum. After all batched materials are together, they should be mixed from 3 to 5 minutes. Less mixing time may result in nonuniformity, poor workability, low water retention, and less than optimum air con-

Fig. 2-12. For best results, mortar should be mixed with a power mortar mixer.

tent. Longer mixing times may adversely affect the air contents of those mortars containing air-entraining cements, particularly during cool or cold weather. Longer mixing times may also reduce the strength of the mortar.

Batching procedures will vary with individual preferences. Experience has shown that good results can be obtained when about three-fourths of the required water, one-half of the sand, and all of the cementitious materials are briefly mixed together. The balance of the sand is then charged and the remaining water added. The amount of water added should be the maximum that can be tolerated and still attain satisfactory workability. Mixing is carried out most effectively when the mixer is charged to its design capacity. Overloading can impair mixing efficiency and mortar uniformity. The mixer drum should be completely empty before charging the next batch.

Mixing by Hand

When hand-mixing of mortar becomes necessary, such as on small jobs, all the dry materials should first be mixed together by hoe, working from one end of a mortar box (or wheelbarrow) and then from the other. Next, two-thirds to three-fourths of the required water is mixed in with the hoe and the mixing continued as above until the batch is uniformly wet. Additional water is carefully added with continued mixing until the desired workability is attained. The batch should be allowed to stand for approximately 5 minutes and then be thoroughly remixed with the hoe.

Retempering

Fresh mortar should be prepared at the rate it is used so that its workability will remain about the same throughout the day. Mortar that has been mixed but not used immediately tends to dry out and stiffen. However, loss of water by absorption and evaporation on a dry day can be reduced by wetting the mortarboard and covering the mortar in the mortar box, wheelbarrow, or tub.

If necessary to restore workability, mortar may be retempered by adding water; thorough remixing is then necessary (Fig. 2-13). Although small additions of water may slightly reduce the compressive strength of the mortar, the end effect is acceptable. Masonry built using a workable plastic mortar has a better bond strength than masonry built using dry, stiff mortar. Mortar used within 1 hour after mixing should not need retempering unless very hot and evaporative ambient conditions prevail.

Mortar that has stiffened because of hydration hardening should be discarded. Since it is difficult to determine by sight or feel whether mortar stiffening is due to evaporation or hydration, the most practical method of determining the suitability of mortar is on the basis of time elapsed after mixing. Mortar should be used

Fig. 2-13. To restore workability, mortar may be retempered.

water and mixing are required at the project. The mortar is available in Types M, S, and N. The standard package sizes, with mortar yield in parentheses, are as follows: 80 lb. ($\frac{2}{3}$ cu.ft.), 40 lb. ($\frac{1}{3}$ cu.ft.), 20 lb. ($\frac{1}{6}$ cu.ft.), and 10 lb. ($\frac{1}{12}$ cu.ft.). Packaged dry mortar is very useful on small jobs, such as projects needing 1 to 5 cu.ft. of mortar, or jobs with limited space to store mortar ingredients. This mortar should meet the requirements of ASTM C387, Specification for Packaged, Dry, Combined Materials for Mortar and Concrete. This document uses property specifications similar to those in Table 2-2.

Dry-Batching

The process of dry-batching all mortar ingredients avoids the need to adjust the mix for moisture content of the sand and ensures consistent portions of sand and cementitious materials. In dry-batching, the cementitious materials and dried sand are accurately weighed and blended at a central plant before delivery to the site in a sealed truck, where the mixture is conveyed into a sealed, weathertight hopper (Fig. 2-14). When the mason contractor is ready for mortar, he has only to draw material from the hopper, add water, and mix. It is apparent that this method offers great convenience and can result in very uniform mortar.

within 2½ hours after mixing to avoid hydration hardening.

Colored mortar is usually not retempered as additional water may cause a significant lightening of the mortar. If necessary, it should be retempered with caution to avoid variations in the color of the hardened mortar.

Mixing During Cold Weather

When masonry construction occurs during periods of freezing weather, facilities should be available and ready for preparing the mortar and protecting fresh masonry work against frost damage. The preparation of mortar for use under these conditions is particularly important. Chapter 5 goes into more detail on these matters.

Special Mortar Techniques

Below are techniques to simplify mortar quality control and batching. These techniques should not be used without the consent of the project engineer.

Packaged-Dry Mortar Materials

Packaged, combined, dry mortar ingredients have been available since 1936. The bag of dry mortar contains cementitious materials and dry sand accurately proportioned and blended at a manufacturing plant. Only

Fig. 2-14. Mortar can be dry-batched at a central plant, delivered in a sealed truck, and stored at the jobside in a weathertight hopper.

Ready Mixed Mortar

Ready mixed mortar is batched at a central location, usually a ready mixed concrete or mortar batch plant, mixed in the plant, and delivered to the jobsite in trowel-ready condition. Ready mixed mortar is made with essentially the same ingredients as conventional mortar, except the mortar contains a special set-controlling (extended-life) admixture that keeps the mortar plastic and workable for a period of more than 2½ hours, usually 24 to 36 hours, but up to 72 hours with certain admixtures.

Some of the advantages in using ready mixed mortar are (1) better uniformity, (2) more consistent workability, (3) less retempering, (4) no need for mixing equipment and mortar ingredients onsite, and (5) no waiting for mortar to be mixed before starting work

Background. The idea of delivering plant- or transit-mixed mortar to the jobsite has been around for years. In the early 1950's, one effort in New York City delivered lime putty and sand to the jobsite where the cement was added. Dry-batching, discussed earlier, was also an early form of ready mixed mortar. Traditionally, the use of retarders, accelerators, and other admixtures has been regarded with reservation by masonry organizations and the design profession. Of primary concern was the effect of admixtures on a mortar's bond strength, workability, durability, compatibility with various masonry units, and the hazards and corrosiveness of the admixtures to workers and buildings. However, admixtures developed in Germany in the early 1970's successfully extended the life of ready mixed mortar without adversely affecting the other mortar properties. Without retarders, mortar has a limited life of about 2½ hours due to cement hydration. This does not warrant the production of large volumes of mortar at any one time as only small quantities are used based on demand and job conditions.

After several years' use in Europe, ready mixed mortar was first used in Canada about 1980 and in the United States in 1982.

Materials and Mixing. Conventional mortar ingredients are combined with a set-controlling admixture and enough water to provide the desired field consistency. The ingredients are mixed at a central location, using either stationary mixers or truck mixers.

An inportant advantage of ready mixed mortar is that the ingredients are accurately measured by weight or by metering devices, thereby producing more uniform mixes closely meeting design specifications. The volumetric methods used in conventionally prepared field mortar can be inaccurate. Accurate ingredient measurement is especially important in colored mortar.

The design of ready mixed mortar is usually based on a performance specification and therefore proper preliminary laboratory testing is required to determine ingredient proporations. Once laboratory proportions

Fig. 2-15. Ready mixed mortar is produced at a central location and delivered in a ready mix truck to the jobsite where it is stored in tubs. A retarding, set-controlling admixture allows the mortar to remain in a workable condition for 24 to 72 hours.

Fig. 2-16. Special 6-cu.-yd. mortar transporter used to delivery ready mixed mortar. Some units can carry more than one type of mortar simultaneously.

are established, they need not be changed throughout the job except for admixture adjustments required to compensate for temperature changes to maintain a constant setting period or life.

Specifications. Ready mixed mortar should meet the requirements of ASTM C1142, Specification for Ready Mixed Mortar for Unit Masonry. The mortar types are designated as Type RM, RS, RN, and RO. Table 2-5 lists the property specifications for these mortars.

Table 2-5. Property Specifications for Ready Mixed Mortar*

Mortar type	Minimum compressive strength, psi at 28 days**	Minimum water retention, %	Maximum air content, %†
RM	2500	75	18
RS	1800	75	18
RN	750	75	18
RO	350	75	18

*Adapted from ASTM C1142.

**The strength values are standard 2-in.-cube strength values. Intermediate values may be specified in accordance with project requirements. Cylindrical specimens (2x4 in. or 3x6 in.) can also be used as long as their strength relationship to cube strength is documented. The 28-day time period starts when the specimens are cast, not when the mortar is initially mixed.

†When structural reinforcement is incorporated in mortar, the maximum air content shall be 12% or bond strength test data shall be provided to justify higher air content.

Delivery and Storage. Trowel-ready mortar is delivered to the jobsite in ready mix trucks, mortar transporters, mortar containers, or hoppers (Figs. 2-15 to 2-18). The mortar is stored in a protected metal or plastic container (¼ to ⅓ cu.yd.) to minimize evaporation and avoid temperature extremes. In some areas, mortar is available in plastic bags or pails (Figs. 2-19 and 2-20). The retarding, set-controlling admixture delays the initial hydration of the cement, causing the mortar to remain plastic and workable for 24 to 36 hours (up to 72 hours in some cases). A rule of thumb that is helpful when ordering mortar is that for concrete block or brick masonry, a mason uses about 0.4 to 0.7 or 0.3 to 0.5 cu.yd. of mortar per day, respectively.

Use. Ready mixed mortar is used the same way as conventional mortar in reinforced and nonreinforced masonry. If the mortar stiffens due to evaporation or absorption of water, the mortar can be retempered once with additional water to restore workability. However, the mortar should not be used beyond its predetermined life expectancy.

When the mortar is placed in between masonry units, the units absorb water from the mortar, thereby removing the set-controlling admixture from solution at which time the ready mixed mortar proceeds to set like normal mortar. Therefore, masonry walls can be constructed at the same rate as walls with normal mortar. Like normal mortar, sufficient water should be present in the mortar to develop proper strength gain; however, special precautions should be taken to reduce evaporation on hot and windy days.

Ready mixed mortar should be used with caution with nonabsorbent units such as glazed units and glass block. Ready mixed mortar should be retarded for no more than 10 hours for nonabsorbent units.

Ready mixed mortar is not a cure-all product. The cautions and concerns of proportioning and using con-

Fig. 2-17. Ready mixed mortar is often delivered in mortar containers (tubs) on a flatbed truck. The mortar-filled containers are unloaded and left at the jobsite.

Fig. 2-18. Mortar hoppers on a flatbed truck delivering ready mixed mortar. The hoppers each have a capacity of about 1⅓ cu.yd.

ventional mortar are also applicable to ready mixed mortar. The basics of using proper ingredients, mix design, and mixing and placing methods used for conventional mortar must be followed. Special information about the use of ready mixed mortar, including jobsite storage requirements, useful life, and allowable extent of retempering, should be provided by the producer.

Tests. Specific controls and procedures must be carefully followed in order to obtain a successful mortar mix with an extended plastic life. Among these are careful mix design and careful batching controls. To maintain quality and consistent behavior of the mor-

Fig. 2-19. Plastic bags containing ready mixed mortar are delivered to the jobsite on pallets. Each bag contains 70 lb. of mortar and 42 bags are needed for one cu.yd.

Fig. 2-20. Five-gallon plastic pails containing ready mixed mortar.

tar, sources of supply for the original mix ingredients should not deviate. Bond and compressive strength, water permeance, and other properties of ready mixed mortar are generally equivalent to the properties of conventional mortar. The applicable laboratory, batch plant, and field quality assurance tests in ASTM C1142 and C780 should be used and a qualified inspector should be present to perform the tests. Specimens for compressive strength tests should be moved from the field to the laboratory after the mortar has undergone final set and the specimens are four or more days old.

Masonry walls built with ready mixed mortar can be loaded at about the same time as walls built with normal mortar. If the loading period is of concern,

ASTM E447 prism tests can be performed to determine proper loading time.

Air Content. The air content should not exceed 18%, as higher air contents may reduce bond strength and increase mortar compressibility; the latter can shorten wall height. Ready mixed mortar producers attempt to keep the air content at or below 18% for the above reasons and above 14% for optimum workability. Mortar compressibility is also related to the mortar's rate of hardening or set time, which in turn is affected by the absorption properties of the masonry units and cold temperatures.

Mobile-Batcher-Mixed Mortar

Mobile-batcher mixers are special trucks mounted with an auger mixer and material holding bins (Fig. 2-21). They batch by volume and continuously mix mortar as the ingredients—cementitious materials, sand, water, and admixtures—are fed into the mixer. The mortar is proportioned and mixed at the jobsite as needed. The trucks can hold enough materials to produce six or more cubic yards of mortar.

With the use of set-controlling (extended-life) admixtures, one truck can produce various extended life mortars onsite and deliver mortar to several projects in one trip. This mortar, which is the same as ready mixed mortar discussed earlier, is then used as needed; when it is depleted in 1 to 3 days, the mobile batcher returns to the job to batch new mortar and refill the mortar tubs. For large or remote projects, the mobile batcher can remain at the jobsite to produce individual smaller quantities of mortar as needed without the set-controlling admixture.

Some of the advantages of ready mixed mortar also apply to mobile-batched mortar. For example, uniform mortar can be consistently batched with this system with little occupation of project space.

Fig. 2-21. Mobile-batcher mixers produce mortar onsite as needed.

Silo-Mixed Mortar

Silo mixers consist of a screw (auger) mixer that is fed dry mortar ingredients from the silo. Single bin silos use preblended mortar ingredients (cementitious materials and dry sand) whereas multicompartment silos house mortar ingredients separately. For example, two compartment silos have one compartment for sand and one compartment for cementitious materials.

The silo is filled with mortar ingredients at a central plant (Fig. 2-22) and delivered to the jobsite by truck and erected (Fig. 2-23). Mortar ingredients are accurately weighed and preblended as needed at the plant to meet specific project requirements. At the project, the unit merely needs to be connected to a pressurized water source and electricity.

A computer controls the mix proportions. The contractor merely pushes a button when mortar is needed and the silo mixer accurately batches the amount of mortar requested. Wireless remote control is also available to save front-end-loader operator time. The mortar is usually placed in a portable tub (Figs. 2-24 and 2-25) or wheelbarrow for easy distribution on the jobsite.

Fig. 2-23. The silo mixer is transported by truck to the jobsite and erected. It is ready to produce mortar after it is connected to electricity and water.

Fig. 2-22. The storage bin/batcher at this central plant stores individual mortar ingredients (masonry cement, portland cement, lime, and sand) in separate compartments. The computer controlled plant batches the dry ingredients by weight and dry blends them prior to discharge into the silos.

Fig. 2-24. Silo mixer producing mortar at a project. This silo mixer can fill the tub with mortar in about 2½ minutes.

Fig. 2-25. After the tub is filled, it is transported by forklift to the masons.

After the silo is empty, it can be replaced with a new full silo or it can be refilled onsite from a bulk tanker truck. The typical silo holds 14 cu.yd. of mortar ingredients, although silos are available in sizes ranging from 14 to 28 cu.yd. For most jobs, one silo provides mortar for 2 to 3 weeks between refills.

Like special techniques discussed earlier, silo mixed mortar can be consistently batched to meet specifications. Mortar is produced as needed, therefore reducing waste. The silo also requires very little jobsite space and is easy to operate.

Modified Mortars

Occasions arise when a modified masonry mortar may be considered beneficial. Modified mortars are conventional masonry mortars altered by either an addition of an admixture at the mixing location or a replacement of one of the basic mortar ingredients. Benefits are appraised from laboratory testing of comparative mortars and testing of walls containing comparative mortars.

ASTM C270 recognizes admixtures when specified, but the present specification does not provide guidance toward accept/reject criteria. The appendix to ASTM C270 provides some guidance and concerns of the masonry industry regarding the indiscriminate use of admixtures.

Modifiers considered for use in the masonry industry are essentially similar to those used in the concrete industry. Concrete technology, however, must be tempered for masonry industry applications and assessment of benefits. Modifiers with varying degrees of acceptance in the masonry industry are classified as follows:

Admixture	Primary benefits
Air-entraining	Freeze-thaw durability, workability
Bonding	Wall tensile strength

Plasticizer	Workability, economy
Set accelerator	Strength increase
Set retarder	Workability retention
Water reducer	Strength, workability
Water repellent	Weather resistance
Pozzolanic	Workability
Color	Esthetic versatility

Selection and use of a mortar modifier should be based on field performance and laboratory testing.

Air-entraining admixtures benefit a modified mortar by reducing water content and increasing workability and freeze-thaw durability. Bond strength may be reduced slightly.

Bonding admixtures increase the adhesion of masonry mortar to masonry units. Organic modifiers provide an air-cure adhesive that increases bond strength of dry masonry. Some bonding agents may experience a strength regression upon wetting.

Plasticizers benefit modified mortar by increasing the workability characteristics of the mortar. Inorganic plasticizers, such as clay, clay-shale, and finely ground limestone, promote workability and water release for cement hydration. Organic plasticizers also promote workability, though mortar may stick more to the mason's tools.

Set accelerators stimulate cement hydration and increase compressive strength of laboratory test mortars at normal temperatures. Under adverse (cold) weather conditions, presumably requiring their use, the benefits must be demonstrated. Chloride and non-chloride accelerators are available, however, the use of chlorides in masonry is not condoned.

Set retarders delay cement hydration providing more time for the mortar to remain usuable without retempering. The benefits are realized primarily during hot weather. Set retarders also are used in marketing ready mixed, set-controlled masonry mortars. The effect of the modifier persists for up to 72 hours but dissipates when retarded mortar contacts absorptive masonry units.

Water reducers lower water demand of masonry mortar which is accompanied by a potential increase in mortar strength. However, masonry mortar in contact with an absorptive masonry unit rapidly looses water to the unit, thus water reducers, singly, may reduce the cement hydration.

Water repellents modify masonry mortar during the early period after construction. If persistent in their performance (not all are), rain penetration of the masonry may be reduced.

Pozzolanic modifiers, because of their fineness and ability to combine with lime, densify mortar and increase strength.

Color modifiers alter the visual appearance of the mortar, either blending or contrasting color of mortar with unit. Color modifiers may affect the physical properties of mortar, so testing should demonstrate the effect of the addition rate.

The manufacturers of mortar modifiers should be consulted to seek test data supporting the manufacturer's claims as to the performance of their products under the anticipated climatic conditions that will prevail during use. Modifiers, be they admixtures or replacements, must not be used indiscriminately.

Estimating Mortar Quantities

It's easy for a beginner to be fooled when estimating the quantity of mortar for a masonry job. The novice will probably estimate too low. If the estimate is used to order the cement and sand for the job, the mason will have to call in for more of the same, and the job may be delayed.

The novice may also be surprised to learn that 0.99 cu.ft. of sand and 0.33 cu.ft. of cementitious material make 1 cu.ft. of mortar.

An estimator might start by calculating how much mortar there is in a typical 8×8×16-in. concrete block wall. It is customary to use only face-shell bedding with ⅜-in.-thick joints. A typical 8-in. concrete block would have 1.25-in.-thick face shells, so the mortar for each block (2 bed and 2 head joints) would amount to:

2×0.375 in. $\times 1.25$ in. $\times 23.62$ in. $= 22.1$ cu. in. per block. There are 112.5 block (8×8×16) in 100 sq.ft. of wall, for which the amount of mortar would be:

$$\frac{22.1 \times 112.5}{1728} = 1.44 \text{ cu.ft.}$$

If one considers that a 10% allowance for waste is generous, the estimator would then order materials for $1.10 \times 1.44 = 1.6$ cu.ft. of mortar for each 100 sq.ft. of wall. What the estimator doesn't know is that most mason contractors order several times as much.

Mortar is a small part of the total cost of a wall, but it is critical to the success of a masonry job. Thus, it does not pay to figure quantities too closely. Following are some practical reasons for allowing plenty of overage when ordering materials for mortar:

Mortar width. The face-shell thickness may be 50% greater on top than the average or minimum thickness quoted.

Droppings. A certain amount of mortar falls to the ground and there is no way to reclaim it.

Accuracy. The mason cannot apply mortar with the precision of a machine.

Sticking. Mortar sticks to the mixer, tub, wheelbarrow, and mortar board.

Dumping. Sand is usually dumped on the ground and can't all be used.

Spilling. Some cement, lime, or mortar is bound to be spilled, spoiled, or lost.

Stopping. Mortar left at the end of the day or at lunchtime must be discarded.

Weather. A sudden turn in the weather causing a job shutdown or spoiled mortar.

Change orders. Misfits of openings, dimensions, etc., discovered at the last minute.

Jambs. Door and window jambs are filled with mortar.

Solid courses. Cores filled solid with mortar for anchor bolts and where joists bear.

First course. The bottom course rests on a generous bedding (¼ to ¾ in. thick).

Cutouts. Plumbing and electrical openings to be closed with mortar.

Tolerances. If all block in a wall were 1/16 in. undersize, all mortar joints would have to be 17% thicker.

Sloughing. Mortar thrown into a wall cavity rather than back onto the mortar board.

Timing. Mortar setting up before being used.

Inspection. Mortar rejected because the mixture was not right.

Judgement. On small jobs, contractors don't calculate quantities, they just "order from experience."

Rounding. When 9.2 cu. yd. of sand are needed, 10 are ordered. When 84 bags of cement are needed, 90 are ordered.

Variations. Sand quality is not always uniform.

Proportioning. Sand quantitity may vary from 2¼ to 3 times the sum of the volume of cementitious material depending on the units and job conditions.

Others. It is not difficult to find other job problems and peculiarities. Human error causes a variety of wastes.

Probably a lot of contractors have forgotten why they order as much mortar as they do. Some of them might say they allow 10% for waste. What they call waste doesn't begin to account for a lot of good reasons for ordering much more than the theoretical quantity. Table 2-6 gives recommendations for the amount of mortar to order for single-wythe walls of various thicknesses using 8-in.-high by 16-in.-long units. For more details on estimating the quantity of mortar and other materials in concrete masonry, see "Quantity Takeoffs," Chapter 6.

Table 2-6. Recommended Mortar Quantities to be Ordered

Nominal wall thickness, in.	Nominal size (width x height x length) of concrete masonry units, in.	Material quantities for 100 sq.ft. of wall area	
		Number of units	Mortar, cu.ft.
4	4x8x16	112.5	8.5
6	6x8x16	112.5	8.5
8	8x8x16	112.5	8.5
12	12x8x16	112.5	8.5

Grout

Grout is an essential element of reinforced brick or concrete masonry (block) construction. The grout bonds the masonry units to the steel reinforcement so they act together to resist loads. Either core holes in the masonry units or the space between wythes are grouted. Reinforced masonry is essential in earthquake zones (Fig. 2-26).

Fig. 2-26. Grout is an essential part of reinforced masonry.

Grouting of brick or block walls serves several purposes: (1) it increases the cross-sectional area of the wall and aids in resisting vertical loads and lateral shear loads, (2) it bonds the wythes together, and (3) it transfers stress from the masonry to the reinforcing steel when a wall is subjected to lateral forces due to wind, earthquake, or earth pressure.

Grouted unreinforced masonry walls are similar to reinforced masonry walls but do not contain reinforcement. Grout is sometimes used in load-bearing wall construction to give added strength to hollow walls by filling a portion or all of the cores. It is also used for filling bond beams and occasionally the collar joint (space between wythes) in two-wythe wall construction.

Grout Selection

Masonry grout is composed of a mixture of cementitious material and aggregate to which sufficient water is added to cause the mixture to flow readily into the masonry cores and cavities without segregation. Unless otherwise specified, grout mix proportions should conform to the requirements of ASTM C476 (Table 2-7). Also, see the discussion on "strength" on the next page for alternate grout design.

The fineness or coarseness of a grout is selected on the basis of the size of the grout space to be filled as well as the height grouted (Table 2-8). Fine and coarse grout and aggregate size and gradation are defined in ASTM C476 and ASTM C404. Building codes and standards sometimes differ on specific values of maximum grout aggregate size versus clear opening, so the governing document should be consulted.

For fine grout (grout without coarse aggregate), the smallest space to be grouted should be at least 3/4 in. wide, as occurs in the collar joint of two-wythe wall construction.

In high-lift grouting where the smallest horizontal dimension of the space to be grouted is about 3 in., a coarse grout with 1/2-in. maximum-size coarse aggregate (or pea gravel) may be used. Some specifying agencies stipulate that 3/4-in. maximum-size coarse aggregate may be used when the grout space is 4 in. or greater. The maximum size of the aggregate and consistency of the mix should be selected considering the particular job conditions to ensure satisfactory place-

Table 2-7. Masonry Grout Proportions by Volume*

Type	Parts by volume of portland cement or blended cement	Parts by volume of hydrated lime or lime putty	Aggregate, measured in a damp loose condition	
			Fine	Coarse
Fine grout	1	0 to 1/10	2¼ to 3 times the sum of the volumes of the cementitious materials	
Coarse grout	1	0 to 1/10	2¼ to 3 times the sum of the volumes of the cementitious materials	1 to 2 times the sum of the volumes of the cementitious materials

*Adapted from ASTM C476.

Notes: Applicable material standards: portland cement—ASTM C150; blended hydraulic cement—ASTM C595; hydrated lime—ASTM C207; lime putty—ASTM C5; aggregate—ASTM C404.

Table 2-8. Maximum Pour Heights and Specified Grout Types with Respect to Grout Space

Maximum grout pour height, ft.	Specified grout type*	Minimum width of grout space, in.**·†	Minimum grout space dimensions for grouting cells of hollow units, in.†·‡
1	Fine	¾	½ by 2
5	Fine	2	2 by 3
12	Fine	2½	2½ by 3
24	Fine	3	3 by 3
1	Coarse	1½	1½ by 3
5	Coarse	2	2½ by 3
12	Coarse	2½	3 by 3
24	Coarse	3¾	3 by 4

*Fine and coarse grouts and aggregates are defined in ASTM C476 and C404.
**For grouting between wythes.
†Grout space dimension equals grout space width minus horizontal reinforcing bar diameter.
‡Area of vertical reinforcement shall not exceed 6% of the area of the grout space.

Adapted from ACI 530.1/ASCE 6 (Ref. 76).

ment of the grout and proper embedment of the reinforcement.

Specifications and Codes

Grout for use in masonry walls should comply with the requirements of ASTM C476 (Table 2-7); Building Code Requirements for Masonry Structures, ACI 530/ASCE 5; and Specifications for Masonry Structures, ACI 530.1/ASCE 6.

Materials

The grout ingredients should meet the requirements of ASTM C150 and C595 for cement, C5 for lime putty, C207 for hydrated lime, and ASTM C404 for aggregates. All of the materials included in ASTM C476 are satisfactory for use in grout. Most projects using large volumes of grout obtain the grout from a ready mixed concrete producer; the use of lime then becomes uneconomical because of the expense in handling.

Admixtures

Practice has shown that a grouting-aid admixture may be desirable when the concrete masonry units are highly absorbent. The desired effect of the grouting aid is to reduce early water loss to the masonry units, to promote bonding of the grout to all interior surfaces of the units, and to produce a slight expansion sufficient to help ensure complete filling of the cavities.

Do not use chloride admixtures in grout because of possible corrosion of reinforcement, metal ties, or an-

chors. *The use of any admixture must be approved by the project engineer.*

Strength

As an alternate to ASTM C476 grout proportions (Table 2-7), grout can be proportioned to have a compressive strength (ASTM C1019) equal to or exceeding the specified compressive strength of the masonry f'_m, but not less than 2,000 psi.

The mix proportions in Table 2-7 will produce grouts with a compressive strength of 1,000 to 2,500 psi at 28 days, depending on the amount of mixing water used, when tested by conventional laboratory methods using nonabsorbtive molds (as are used for sampling and testing mortar and concrete). However, the actual in-place compressive strength of grout generally will exceed 2,500 psi because, under ordinary conditions, some of the mixing water will be absorbed by the masonry during the time the grout is placed and prior to setting and hardening. This absorption of moisture, in effect, reduces the water-cement ratio of the in-place grout and increases the compressive strength. The moisture absorbed and held by the surrounding masonry during the period immediately following placement of grout helps to maintain the grout in the moist condition needed for satisfactory cement hydration and strength gain.

Consistency

All grout should be of a fluid consistency but only fluid enough to pour or pump without segregation. It should flow readily around the reinforcing steel and into all joints of the masonry, leaving no voids. There should be no bridging or honeycombing of the grout.

The consistency of the grout as measured using a slump test (ASTM C143) should be based on the rate of absorption of the masonry units and on temperature and humidity conditions. The slump should be between 8 and 11 in. (Fig. 2-27)—about 8 in. for units with low absorption and about 10 in. for units with high absorption.

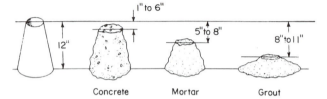

Fig. 2-27. Slump test comparison of concrete, mortar, and masonry grout.

Fig. 2-28. Pumping grout.

Mixing

Whenever possible, grout should be batched, mixed, and delivered in accordance with the requirements for ready mixed concrete (ASTM C94). Because of its high slump, ready mixed grout should be continuously agitated after mixing and until placement to prevent segregation.

Mixing of masonry grout on the jobsite is usually not recommended unless unusual conditions exist. When a batch mixer is used on the jobsite, all materials should be mixed thoroughly for a minimum of 5 minutes. Grout not placed within 1½ hours after water is first added to the batch should be discarded.

Placing

Even though masonry grout is quite fluid, it is a good practice to consolidate the grout by rodding or vibration to ensure that it encompasses all the reinforcing steel and completely fills the voids. Grout pours up to 12 in. high are consolidated by vibration or rodding (puddling). Grout pours more than 12 in. high are consolidated by vibration and reconsolidated after settlement and initial water loss occurs.

Because grout mix water is absorbed by the masonry units, there is a slight volume reduction of the grout. Therefore, the use of shrinkage-compensating admixtures or expansive cement is sometimes recommended on highlift grouting construction. The expansion of this type of grout counteracts the volume change due to loss of water to the masonry units.

Masonry grout is usually delivered in a truck mixer and pumps are used to place the grout in the walls (Fig. 2-28). The grout is placed in lifts up to 5 ft. deep.

Curing

The high water content of masonry grout and the partial absorption of this water by the masonry units will generally provide adequate moisture within the masonry for curing both the mortar and grout. In dry areas where the masonry is subjected to high winds, some moist curing (such as fogging or protection with plastic sheeting) may be necessary. Grout placed during cold weather is particularly vulnerable to freezing during the early period after grouting because of its high water content. To offset cold temperature, grout sand and water can be heated and heated enclosures or covers can be used to protect masonry when temperatures are below 40°F.

Sampling and Testing

ASTM standard C1019 can be used for quality control of uniformity of grout during construction or as an aid in helping to select grout proportions. For making compressive test specimens the standard uses molds formed with masonry units having the same absorption characteristics and moisture content as the units used in the construction (Fig. 2-29). This simulates in-the-structure conditions, where water from the grout is freely absorbed by the units, thus reducing the water-cement ratio of the grout and increasing its strength. Grout specimens for compressive tests should be prepared according to ASTM C1019 rather than cast in the usual cylinder molds used for concrete samples because the high water content of the grout would cause low strength results not indicative of actual in-the-wall strength. A mimimum of ½ cu.ft. of grout

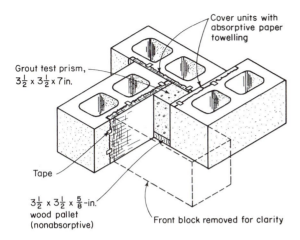

Cover units with
absorptive paper
towelling

Grout test prism,
$3\frac{1}{2}$ x $3\frac{1}{2}$ x 7 in.

Tape

$3\frac{1}{2}$ x $3\frac{1}{2}$ x $\frac{5}{8}$-in.
wood pallet
(nonabsorptive)

Front block removed for clarity

Mold with four 8 x 8 x 16-in. blocks

Fig. 2-29. ASTM C1019 method of using masonry units to form a prism for compression-testing of masonry grout.

should be sampled for slump and strength tests. The grout sample should be obtained as the grout is placed into the wall. The specimen should be a nominal 3 in. or larger square prism with a height of twice the width. Three test prisms should represent one grout sample.

When required, grout samples should be taken and tested as per ASTM C1019 for each 5,000 sq.ft. of masonry. Also, a sample should be taken whenever there is any change in mix proportions, method of mixing, or materials used.

The duties and responsibilities of testing agencies are described in ASTM practice C1093.

Chapter 3

PROPERTIES OF CONCRETE MASONRY WALLS

A wall does more than merely enclose a building in an attractive fashion. It must have strength to support floors and roofs and to resist the buffeting of nature. It must be a shield against noise, heat or cold, and fire damage. This chapter assists architects, engineers, and builders in the design and construction of concrete masonry walls that fill these needs.

Strength and Structural Stability

Modern concrete masonry wall construction is of two general types: unreinforced (plain) and reinforced. These classifications are characterized by some differences in mortar type requirements, use of reinforcing steel, and erection techniques. Both types are usually subject to the provisions of applicable building codes.

Unreinforced (plain) concrete masonry is the ordinary type that has been in use for many years. Essentially unreinforced, any steel reinforcement used in this type of concrete masonry is generally of light gage and placed in relatively small quantities in the horizontal joints.

Reinforced concrete masonry on the other hand contains reinforcing steel so placed and embedded that the masonry and steel act together in resisting forces. This structural behavior is obtained by placing deformed reinforcing steel bars in continuous vertical and horizontal cores or cavities in the masonry and then filling these spaces with properly consolidated portland cement grout. Structural bond develops between the hardened grout, the bars, and the masonry units, permitting reinforced concrete design theory to be adapted to produce buildings of reinforced concrete masonry.

Reinforced concrete masonry is used where the compressive, flexural, and shear loads are higher than can be accommodated with plain concrete masonry. It is required by code in areas of recurring hurricane winds or earthquake activity where major damage to buildings is highly probable.

Structural Tests

Extensive testing of representative concrete masonry assemblages over a period of many years has established concrete masonry wall construction as a reliable and predictable structural system. Some of the test methods developed during structural research studies* of concrete masonry and other wall forms have been adopted as standards and are described in ASTM E72, Standard Methods of Conducting Strength Tests of Panels for Building Construction. Some other structural tests that are commonly used in analyzing masonry include ASTM C952, C1006, and C1072.

Analyses of accumulated test data have produced useful, reproducible relationships between the strengths of masonry components and the strengths of completed walls. From these relationships building code authorities have established allowable design stresses under various loading conditions or formulas for their calculation. Furthermore, structural research has enabled the development of quality control criteria to ensure fulfillment of design requirements in the completed structure.

Design Methods

Design methods for concrete masonry are well established. However, there is continuing research to insure that these methods are updated to keep pace with progress in the construction industry.

---

*See Ref. 40.

Fig. 3-1. Reinforced concrete masonry. Steel is placed in the vertical and horizontal cavities before they are filled with grout.

Fig. 3-2. High-rise load-bearing walls with round fluted concrete masonry units.

Fig. 3-3. Concrete masonry wall being tested for flexural strength at the University of Illinois.

Fig. 3-4. A concrete masonry research project in the structural research laboratory of the Portland Cement Association. Specimens are being investigated for shrinkage and creep.

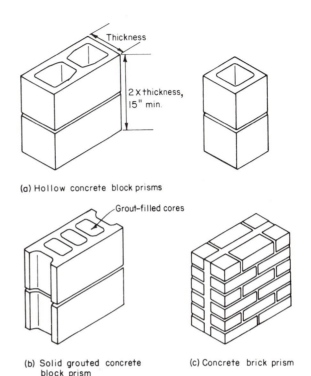

(a) Hollow concrete block prisms

(b) Solid grouted concrete block prism

(c) Concrete brick prism

Fig. 3-5. Concrete masonry compression-test prisms.

In 1988, a joint committee of the American Concrete Institute (ACI) and the American Society of Civil Engineers (ASCE) issued *Building Code Requirements for Masonry Structures* (ACI 530-88/ASCE 5-88), including commentary (see References 75 and 77). This new Code reflects significant advances in design requirements and procedures. It is based on allowable or working stress design methods. Design provisions that allow or neglect masonry tensile stress are included. Under certain limitations, and for buildings of smaller scale, empirical design provisions are also included in the new Code.

Strength design methods based on limit state using load and strength reduction factors are making their way into masonry building codes and design manuals. The 1988 *Uniform Building Code* (Reference 81) already permits the design of concrete masonry slender walls and shear walls to be based on these methods. A subcommittee of ACI Committee 530, Masonry Structures, is working on recommendations to introduce strength design requirements into coming editions of *Building Code Requirements for Masonry Structures*.

The use of strength design represents an advance in the way masonry structures are engineered. It should be noted that strength design calculations should be coupled with calculations to check structural serviceability requirements, such as deflection and crack control. Use of strength design in masonry also will facilitate the work of engineers involved in a total project since reinforced concrete and structural steel already employ strength design. More information on design methods can be found in References 81 and 97.

Basic Concepts

In the design of concrete masonry, the compressive strength of the masonry is a basic quantity. The compressive strength of masonry must equal or exceed the specified compressive strength of masonry, f'_m, used in the structural design. Allowable stresses in masonry are a percentage of the specified masonry compressive strength, f'_m, often expressed in pounds per square inch. The compressive strength depends on the strength of the masonry units and mortar in the joints and cavities. According to ACI 530.1, two methods may be used to determine masonry strength for the particular type of mortar and masonry units to be used on the project:

1. Masonry prisms representative of the masonry wall are constructed as shown in Fig. 3-5. If the structure is to have grouted cores, the masonry unit cores in the prism are filled with grout. No reinforcement is used in the prisms, but a two-wythe prism may contain metal ties. Test procedure, apparatus, and data should comply with requirements of ASTM E447. The compressive strength is calculated by dividing the ultimate test load by the solid cross-sectional area of the prism, including the area of mortar filled cavities. Additional details are given in the commentary to Chapter 24 of Reference 81 and in ASTM E447.

2. If no prism tests are conducted, compressive strength may be assumed based on tables given in building codes and specifications (such as Ta-

Table 3-1. Compressive Strength of Masonry Based on Compressive Strength of Concrete Masonry Units and Type of Mortar Used in Construction*

Net area compressive strength of concrete masonry units, psi		Net area compressive strength of masonry, psi**
Type M or S mortar	Type N mortar	
1,250	1,300	1,000
1,900	2,150	1,500
2,800	3,050	2,000
3,750	4,050	2,500
4,800	5,250	3,000

*Adapted from Ref. 76.
**For units less than 4 in. high, use 85% of the values listed.

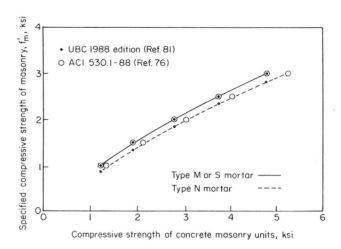

Fig. 3-6. Value of f'_m based on strength of individual concrete masonry units.

ble 3-1 adapted from Reference 76). The relationship between masonry-unit strength and specified masonry compressive strength is illustrated in Fig. 3-6.

The value of f'_m may be assumed to be 75% of the average calculated value of previous tests conducted on prisms constructed using the same materials in the same conditions as indicated in the commentary to Chapter 24 of Reference 81.

Tests of masonry prisms may be considered conceptually analogous to concrete cylinder tests in concrete construction. In concrete, the variability of test results is taken into consideraiton based on criteria using standard deviation calculations. For masonry, it is not possible at this time to provide similar criteria because of lack of sufficient test data. Currently, the compressive strength is determined as the average of test results of at least three masonry prisms. This average strength must equal or exceed f'_m.

Since strength design is already permitted by the *Uniform Building Code*, basic concepts concerning this method will be reviewed.

In working stress design, calculations are performed to determine stresses in a member under the action of service loads prescribed in building codes. Calculated stresses are limited to allowable levels of stress prescribed by the codes. The calculations are based on the assumption of material behavior following a linear relationship between stress and strain. For a purely flexural member, such as a masonry lintel, the distribution of strain and stress under the prescribed service load is assumed to be as shown in Fig. 3-7a.

In strength design, calculations are performed to determine the strength of the member under the action of factored loads obtained by multiplying code prescribed service loads by code prescribed load factors. Under factored loads, material within the member is assumed to be at the point of failure. For example, concrete masonry is assumed to be at the point of crushing and steel at the point of yield. To account for

conditions that include variability in material strength and tolerances in member dimensions, calculated strength is multiplied by strength reduction factors prescribed by code.

It should be noted that while calculations are performed assuming conditions at failure, code prescribed load factors and strength reduction factors provide the safety margin intended by code for the design of the member. For a purely flexural member, such as a masonry lintel, the distribution of strain and stress is assumed to be as shown in Fig. 3-7b.

Under both working stress and strength design methods, calculations must be performed to check serviceability conditions, such as estimating deflections and evaluating the possibility of excessive cracking. In Fig. 3-7b, a rectangular stress distribution is used to reflect the nonlinearity of stress and strain in concrete or masonry near failure. Stress distribution and assumptions used for concrete masonry are currently the same as those used for concrete. Additional research may yield a revision of these assumptions for concrete masonry.

Effect of Loads

The type of load acting on a masonry wall affects its design and proportioning. The different types are axial loads, lateral loads, or shear loads as illustrated in Fig. 3-8.

Axial Loads. The allowable concentric axial load on an unreinforced laterally supported concrete masonry wall depends on such features as wall height, wall thickness, and masonry compressive strength (see commentary to Chapter 24 of Reference 81). To facilitate selection of proper wall dimensions, a variety of design aids such as Table 3-2 are available.

It is assumed for all the allowable wall loads shown in the tables and graphs of this section that proper

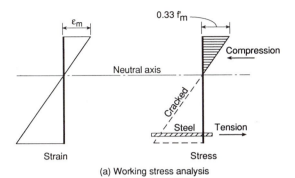

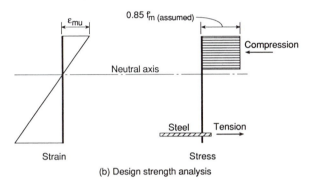

Fig. 3-7. Strain and stress distributions at cross section of a concrete masonry flexural element.

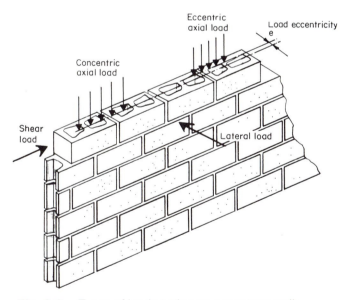

Fig. 3-8. Types of loads acting on a masonry wall.

Table 3-2. Allowable Axial Loads on Single-Wythe, Hollow Concrete Masonry Walls*

Height of wall, ft.	Allowable axial loads, lb. per lin.ft.			
	$t = 6$ in.	8	10	12
	$A_e = 40.5$ in.^{2}/ft.	45	61	70
6	10,680	12,030	16,387	18,845
7	10,935	11,960	16,338	18,813
8	10,330	11,867	16,273	18,769
9	10,074	11,747	16,190	18,714
10	9,754	11,597	16,086	18,645
11		11,413	15,959	18,560
12		11,194	15,806	18,459
13		10,934	15,626	18,340
14			15,416	18,200
15			15,174	18,039
16			14,897	17,855
17				17,647
18				17,412
19				17,150
20				16,859

*Adapted from Plate 4-16S of Ref. 59, and Ref. 81.
Notes:
1. Self weight of wall not included.
2. Wall is braced at the top.
3. Allowable load (N) is based on formula:
$$N = 0.2 f'_m[1-(h/42t)^3]A_e \text{ (Ref. 81)}$$
where f'_m = specified compressive strength = 1350 psi
     h = wall height (in.)
     t = wall thickness (in.)
     A_e = effective area (sq.in./lin.ft.)
4. A_e values are adapted from Ref. 20, assuming masonry units have fully bedded joints.

Vertical loads on load-bearing walls include the weight of floors, walls, and roof above, as well as the weight of the wall itself. When vertical loads are not centered on the wall, additional stresses due to bending are created. The combination of axial and bending stresses can be calculated and compared with allowable stresses given in building codes. For eccentric loads, design aids such as Table 3-3 may be used to facilitate selection of wall dimensions (also see References 12 and 20).

Lateral Loads. Lateral loads induce significant bending stresses in masonry walls. Basement walls are subject to lateral soil pressures. Also, if drainage has not been provided around the base of the foundation, hydrostatic pressure at the wall surface increases bending stresses in the wall. Table 3-4 shows required thicknesses of foundation walls subject to lateral soil pressure. Note that in most cases basement walls with a masonry superstructure above can sustain a greater depth of unbalanced fill than walls with a wood frame above. In other words, the wall does not have to be as thick with a masonry wall above as with a wood frame above; this is because the heavier superimposed load conteracts the tensile bending stresses to a greater extent.

engineering or architectural supervision of construction (quality assurance) will take place. If this will not occur, the design codes specify that the loads must be drastically reduced. Inspection of engineered concrete masonry is necessary if its economic potential is to be realized.

Table 3-3. Allowable Loads When Maximum Load is Governed by Tension Due to Eccentricity of Load on Nonreinforced Concrete Masonry Walls*

	8-in wall			10-in wall			12-in wall	
	Allowable load, kips/ft.			Allowable load, kips/ft.			Allowable load, kips/ft.	
e, in.	Type M & S mortar	Type N mortar	e, in.	Type M & S mortar	Type N mortar	e, in.	Type M & S mortar	Type N mortar
2.00	2.74	1.90	2.25	3.97	2.76	3.00	4.60	3.20
2.20	2.11	1.47	2.50	2.95	2.05	3.25	3.59	2.50
2.40	1.72	1.19	2.75	2.35	1.63	3.50	2.99	2.08
2.60	1.45	1.01	3.00	1.95	1.36	3.75	2.56	1.78
2.80	1.25	0.87	3.25	1.67	1.16	4.00	2.23	1.55
3.00	1.10	0.77	3.50	1.46	1.01	4.25	1.98	1.38
3.20	0.98	0.68	3.75	1.29	0.90	4.50	1.78	1.24
3.40	0.89	0.62	4.00	1.16	0.81	4.75	1.61	1.12
3.60	0.81	0.56	4.25	1.06	0.73	5.00	1.47	1.03
3.80	0.74	0.52	4.50	0.97	0.67	5.25	1.36	0.95

*Adapted from Ref. 21.

Notes:
Design is based on NCMA criteria: $f_m - 0.75f_a \leq F_t$

where f_m = computed flexural tensile stress
f_a = computed axial stress
F_t = allowable tensile stress; 23 psi for Type M & S mortar, 16 psi for Type N mortar.
(In the new 1988 ACI masonry code, these allowable stresses are slightly higher.)

Table 3-4. Minimum Thickness and Allowable Maximum Depth of Unbalanced Fill for Nonreinforced Concrete Masonry Basement Walls*

Type of basement wall construction	Minimum nominal thickness, in.	Maximum depth of unbalanced fill, ft.**		
		Type of superstructure		
		Wood frame	Masonry veneer	Masonry
Hollow concrete masonry units	8	4	4.5	5
	10	5	5.5	6
	12	7	7	7
Solid concrete masonry units	6	3	4	4
	8	5	5.5	6
	10	6	6	6.5
	12	7	7	7

*Adapted from Ref. 64.
**Unbalanced fill is the height of outside finish grade above the basement floor or inside grade.

Wind and earthquakes also cause lateral forces on masonry walls (see References 68 and 81). Wind pressures are prescribed by building codes and range from 5 to 30 psf.

In seismic regions building codes call for lateral loads ranging from 5% to 20% of the dead weight of the wall. While wind loads are applicable for exterior walls, earthquake loads act on both exterior and interior walls.

Concrete masonry walls resist lateral loads in the horizontal and vertical spans, depending on the height and length between cross walls. When the distance between lateral supports exceeds the height of the wall, bending due to wind is principally in the vertical span.

This is illustrated in Fig. 3-9; the diagram represents a typical one-story wall bounded by laterally supported, vertical control joints and having no windows or doors.

For wind, an example of allowable height of a one-story wall spanning vertically is given in Fig. 3-10. Allowable horizontal span of the same type of wall is given in Fig. 3-11.

If a wall resists wind in its vertical span, a concentric compressive vertical load on the wall will increase its strength (up to a point). For example, in Fig. 3-10, a 12-in. wall has an allowable height of 16 ft. against a wind pressure of 20 psf. However, Table 3-5 shows that as the axial load increases on this same wall, its allowable height increases. On the other hand, when a wall is subject to a vertical load but resists wind by its horizontal span, the two effects are not additive.

Reinforcing bars increase the strength of concrete masonry walls. This can be seen by comparing a reinforced wall in Table 3-6 with a similar but unreinforced wall in Table 3-5.

The 1988 *Uniform Building Code* permits the design of concrete masonry walls subject to combined axial and lateral loads to be based on strength design methods. As indicted in Reference 98, this approach leads to economical designs. In addition, the design of slender walls can be handled taking into account secondary effects due to lateral deflection of the wall. When using strength design construction, quality control is essential to ensure proper installation of details shown in

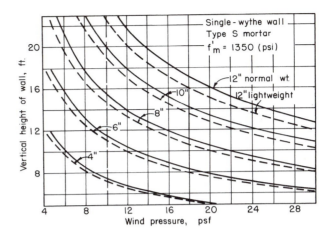

Fig. 3-10. Allowable height of non-load-bearing concrete masonry single-wythe walls for wind. (Adapted from Plate 4-11S, Ref. 59.)

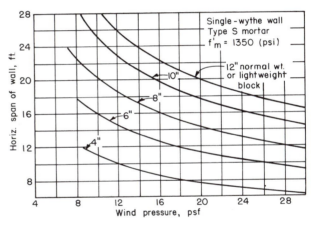

Fig. 3-11. Allowable horizontal span of non-load-bearing concrete masonry single-wythe walls for wind. (Adapted from Plate 4-12S, Ref. 59.)

design drawings. Designers should also recognize the importance of details such as reinforcement anchorage and proper placement of mortar both within the masonry unit cavities and at the joints.

Masonry walls should have adequate lateral support to resist both axial and lateral loads. The 1988 ACI *Building Code Requirements for Masonry Structures* (see empirical design provisions in Reference 75) recommends maximum wall height or maximum distance between lateral supports for a given wall thickness.

Shear Loads. Wind and earthquakes induce out-of-plane pressure as well as in-plane forces or shear loads on masonry walls. In combination with roof diaphragms, concrete masonry walls are used effectively to provide lateral stability for buildings under wind and earthquake loads. The 1988 *Uniform Building*

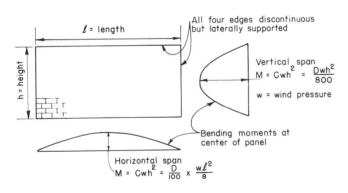

Vertical span
$$M = Cwh^2 = \frac{Dwh^2}{800}$$

w = wind pressure

Bending moments at center of panel

Horizontal span
$$M = Cwh^2 = \frac{D}{100} \times \frac{w\ell^2}{8}$$

ℓ/h	Bending moment coefficient (C)		Virtual percent of wind creating bending (D)	
	Vertical span	Horizontal span	Vertical span	Horizontal span
1	0.044	0.044	35	35
1½	0.078	0.043	62	15
2	0.100	0.037	80	7
2½	0.112	0.032	90	4
3	0.118	0.029	95	3

Fig. 3-9. Distribution of wind forces on a wall panel.

us^^^

Table 3-5. Maximum Wall Heights for Nonreinforced Hollow Unit Concrete Masonry Walls Subjected to Concentric Loads and Wind*

| Concentric load, plf | Maximum wall heights, ft.** | | | | | | | | | | | |
| | t = 6 in. | | | 8 in. | | | 10 in. | | | 12 in. | | |
	w = 10	20	30	10	20	30	10	20	30	10	20	30
0	10.0	7.1	5.8	12.8	9.0	7.4	16.0	11.3	9.2	19.6	13.9	11.4
500		8.4	6.9	13.3	10.5	8.6	16.7	12.7	10.4	20.0	15.4	12.6
1,000		9.6	7.8		11.9	9.7		14.0	11.4		16.8	13.8
1,500		10.0	8.7		13.0	10.6		15.2	12.4		18.2	14.8
2,000			9.4		13.3	11.5		16.2	13.3		19.3	15.8
2,500			10.0			12.4		16.7	14.1		20.0	16.7
3,000						13.1			14.9			17.6
3,500						13.3			15.6			18.4
4,000									16.3			19.2
4,500									16.7			20.0

*Design assumptions: Minimum f'_m = 1350 psi; Type S mortar, allowable flexural tensile stress = 23 × 1.33 = 30.6 psi; w = wind load, psf.
**Maximum wall heights limited to 20 times nominal wall thickness (t).
Adapted from Ref. 38

Table 3-6. Allowable Vertical Axial Loads for Reinforced Hollow Unit Concrete Masonry Walls Subject to Wind*

| Wall height, ft. | Allowable vertical axial load, kips per lin.ft. | | | | | | | | | | | |
| | 6-in. wall with No. 4 bars | | | 8-in. wall with No. 5 bars | | | 10-in. wall with No. 6 bars | | | 12-in. wall with No. 7 bars | | |
	w = 10	20	30	10	20	30	10	20	30	10	20	30
8	12.6	9.9	7.3	18.0	16.3	14.5	23.1	21.8	20.5	27.8	26.8	25.8
10	10.4	7.5	2.6	16.6	13.9	11.1	22.1	20.0	18.0	27.1	25.5	23.9
12	7.8	2.7		14.8	11.0	7.2	20.7	17.9	15.0	26.0	23.7	21.5
14	5.0			12.6	7.8	2.9	19.1	15.3	11.5	24.8	21.8	18.8
15	3.6			11.0	5.9		18.0	13.8	9.5	24.0	20.5	17.1
16				10.2	4.4		17.3	12.5	7.7	23.2	19.4	15.5
18				7.7			15.2	9.4	3.7	21.7	16.9	12.2
20				5.1			12.8	6.3		19.5	13.9	8.3
22							10.4	3.2		17.8	11.2	4.7
24							7.9			15.1	8.0	
25							6.7			13.9	6.5	
26										13.2	5.3	
30										8.4		

*Design assumptions: f'_m = 1,500 psi, reinforcing bars are centered on wall at 32 in. oc, w = wind load, psf.
Adapted from Ref. 38

Code permits the design of concrete masonry shear walls using working stress and strength design methods. Reinforcement can be used in mortar filled cavities to provide resistance to overturning forces in the plane of the wall. Reference 98 provides addditional information on the design of concrete masonry shear walls.

Cavity Walls. Cavity walls have different strength characteristics than single-wythe walls. Wind pressure is resisted by non-load-bearing cavity walls as shown in Figs. 3-12 and 3-13. These charts are for walls resisting wind in the vertical and horizontal spans, respectively.

Each wythe in a cavity wall helps resist wind by acting as a separate wall. If both wythes have the same thickness, they resist the wind equally but they do not act as if fully bonded together. For example, in Fig. 3-12 a 10-in. cavity wall (labled 4″-2″-4″) exposed to a 10-psf wind has an allowable height of 11 ft.; but, as shown in Fig. 3-10 a 10-in. single-wythe wall of the same

Table 3-7. Width of Face Shell in Hollow Concrete Masonry Units*

Nominal wall thickness, in.	4	6	8	10	12
Width of face shell, in.	1	1.19	1.25	1.50	1.50

*Adapted from Ref. 59.

Table 3-8. Allowable Shear and Moment on Lintels*

Steel rebar	V_{Rv}, kips	V_{Ru}, kips	M_R, kip-ft.	V_{Rv}, kips	V_{Ru}, kips	M_R, kip-ft.
Bottom cover		1.5 in.			2.0 in.	
8 in. lintel, 8 in. high						
1 # 3	1.829	1.017	0.558†	1.675	0.928	0.904**
1 # 4	1.810	1.304	1.524	1.656	1.187	1.315
2 # 3	1.829	1.967	1.603	1.675	1.793	1.383
1 # 5	1.790	1.572	1.726	1.636	1.429	1.481
2 # 4	1.810	2.505	1.897	1.656	2.278	1.628
1 # 6	1.771	1.824	1.878	1.617	1.656	1.604
1 # 7	1.752	2.061	2.000	1.598	1.868	1.703
2 # 5	1.790	3.006	2.095	1.637	2.730	1.789
1 # 8	1.733	2.284	2.095	1.579	2.067	1.775
2 # 6	1.771	3.479	2.232	1.617	3.156	1.896
2 # 7	1.752	3.924	2.330	1.598	3.556	1.972
2 # 8	1.733	4.346	2.397	1.579	3.934	2.019
8 in. lintel, 16 in. high						
1 # 3	4.293	2.462	1.799†	4.139	2.370	1.733†
1 # 4	4.273	3.201	3.190	4.120	3.081	3.070†
2 # 3	4.293	4.805	3.512†	4.139	4.626	3.381†
1 # 5	4.254	3.908	6.441**	4.100	3.760	6.198**
2 # 4	4.273	6.209	8.020	4.120	5.974	7.550
1 # 6	4.235	4.590	8.174	4.081	4.414	7.686
1 # 7	4.216	5.241	8.998	4.062	5.039	8.452
2 # 5	4.254	7.543	9.231	4.100	7.255	8.677
1 # 8	4.196	5.866	9.724	4.042	5.638	9.125
2 # 6	4.235	8.824	10.198	4.081	8.483	9.578
2 # 7	4.216	10.043	11.041	4.062	9.652	10.349
2 # 8	4.196	11.211	11.750	4.042	10.771	11.002

*Adapted from Ref. 99. See Ref. 99 for other size lintels.
Notes:
$f'_m = 1,350$ psi
V_{Rv} = shear capacity in diagonal tension
V_{Ru} = shear capacity in flexural bond
M_R = moment capacity with grade 40 bars
**Indicates that M_R is governed by tensile stress in steel
†Indicates that $p < 100/F_y$, and M_R is accordingly reduced 25%

quality materials and exposed to a 10-psf wind has an allowable height of 20 ft. This illustrates that cavity walls have less strength. However, strength is often not a deciding factor and cavity walls have distinct advantages: for example, to keep water out and provide good insulation.

Vertical loads on cavity walls are often carried only by the inner wythe. Fig. 3-14 shows some allowable

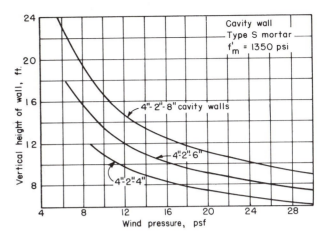

Fig. 3-12. Allowable height of non-load-bearing concrete masonry cavity walls for wind. (Adapted from Plate 6-4S, Ref. 59.)

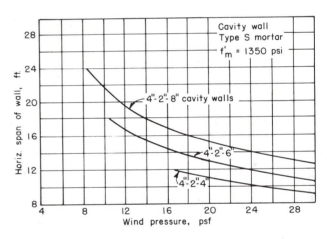

Fig. 3-13. Allowable horizontal span of non-load-bearing concrete masonry cavity walls for wind. (Adapted from Plate 6-5S, Ref. 59.)

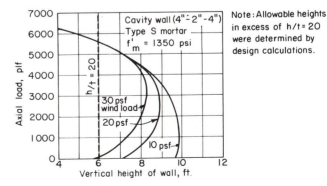

Fig. 3-14. Allowable load and vertical height of concrete masonry cavity walls for wind. Both wythes are 4 in. and vertical load is centered on the inside wythe. (Adapted from Plate 6-6A, Ref. 59.)

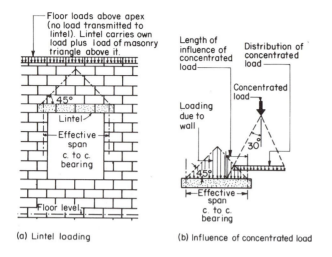

Fig. 3-15. Loads supported by lintels.

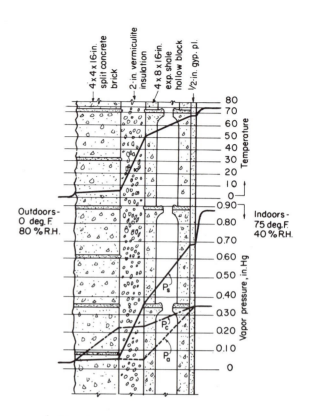

Fig. 3-16. Temperature and vapor pressure gradients for insulated concrete masonry cavity wall construction ($R = 7.26$, $U = 0.14$).

loads and heights for this condition where horizontal span is not involved.

Figs. 3-10 through 3-14 are illustrative examples, based on Ref. 59, for hollow concrete masonry walls with mortar bed widths as given in Table 3-7. Bending moments are computed for simple spans.

Lintels. Above each window or door opening in a wall, there should be a structural beam (lintel) to carry the wall loads, as shown in Fig. 3-15. Table 3-8 gives data for typical reinforced concrete masonry lintels, according to size and reinforcing bars. Deflection of lintels usually is not a problem. However, deflection should be calculated and limited to 1/600 of the clear span for spans up to 15 ft., and 0.3 in. for spans 15 ft. and over. Lintel design is discussed further in the next chapter.

Thermal Insulation

In selecting and sizing the heating and cooling equipment for a building, the designer must first determine the heating and cooling loads involved. These loads are made up of various exchanges of heat, including transfer of heat through exterior walls.

When the outdoor temperature is below the established indoor temperature, heat is lost from the building and the heating system must replace it. Conversely, when the outdoor temperature is above the indoor temperature, heat is gained. In air-conditioned buildings this heat is absorbed by the cooling system. The upper part of Fig. 3-16 gives an example of heat flow through a concrete masonry wall. The temperature gradient varies form 75 deg. F. inside to 0 deg. F. outside.

U and R Values Defined

Essential to the designer's calculations regarding the flow of heat through walls as well as other building components are U values, the coefficients or indices of total heat flow rate. They express the total amount of heat in British thermal units (Btu) that 1 sq.ft. of wall (or ceiling or floor) will transmit per hour for each degree Fahrenheit of temperature difference between the air on the warm and cool sides.

Another index of heat transfer is the R value, which is a measure of the resistance that a building section, material, and air space or surface film offers to the flow of heat. The R value is the reciprocal of a heat transfer coefficient such as U or f. The R value is for a *stated* thickness of a building section, and the unit of value for R is ($°F \cdot h \cdot ft^2$)/Btu. R values are particularly useful for estimating the effect of components of a section on the total heat flow because they can be directly added.

The overall heat transmission coefficient of the wall, the U value, includes the effect of air film or surface conductance for the inside of the wall, f_i, as well as for its outside, f_o. The U value is calculated by taking the reciprocal of this total: $1/f_i + 1/f_o +$ the sum of the R

values for each component of the wall.* *U* values cannot be added or subtracted with meaningful results.

The *U* value for a wall section can be determined by test (ASTM C236, Test Method for Steady-State Thermal Performance of Building Assemblies by Means of a Guarded Hot Box), and many wall tests have been conducted for this purpose over the years. Actually, the rational method for estimating the *U* value by calculation was devised because of the innumerable types and combinations of materials going into modern building walls and the continuing development of new types. The method is based on the heat-flow resistance concept and principles relating to the flow of electricity. Heat-flow resistance is analogous to electrical resistance. Thus, the flow of heat is directly proportional to the temperature difference and inversely proportional to the heat resistance.

Selecting the *U* Value

The hourly heat flow (Btu of heat loss or gain) through the exterior wall construction of a building can be determined by multiplying the *U* value of the section by the total net area in square feet (with areas of windows and doors deducted and dealt with separately) and then multiplying that answer by the design temperature difference between the inside and the outside air. Since the surface area and temperature difference are generally established or fixed quantities, the *U* value remains alone as subject to design adjustment.

In selecting the proper *U* value for a wall section, the designer will consider and weigh several factors. The relative importance of each will depend upon the type of building and its occupancy. For buildings intended for residential use—houses, apartments, dormitories, hotels, etc.—human comfort needs probably will govern the selection. For commercial buildings devoted to manufacturing operations, process control requirements may be the primary factor in deciding the *U* value. In all cases, operational costs, that is, the cost of fuel or electric power, will require careful consideration.

The requirements for building insulation have been undergoing a thorough change in recent times due to the emphasis on conservation of energy and fuels for winter heating and summer cooling. Since national standards will probably continue to evolve, no attempt will be made here to quote insulation requirements of any specific regulatory agencies. The reader is referred to releases from such organizations as the local building department, the U.S. Department of Housing and Urban Development, and the American Society of Heating, Refrigerating and Air-Conditioning Engineers, Inc.

Typical Values for Concrete Masonry Walls

Concrete masonry walls offer insulation qualities combined with architectural appeal. Table 3-9 lists *R* values for single-wythe walls and Table 3-10 gives *R* values for

*See Refs. 22, 29 and 32.

Table 3-9. Heat Resistance (R) Values of Single-Wythe Concrete Masonry Walls*

Nominal wall thickness, in,	Insulation in cells**	R value based on concrete unit weight				
		60 pcf	80 pcf	100 pcf	120 pcf	140 pcf
4	Filled	3.36	2.79	2.33	1.92	1.14
	Empty	2.07	1.68	1.40	1.17	0.77
6	Filled	5.59	4.59	3.72	2.95	1.59
	Empty	2.25	1.83	1.53	1.29	0.86
8	Filled	7.46	6.06	4.85	3.79	1.98
	Empty	2.30	2.12	1.75	1.46	0.98
10	Filled	9.35	7.45	5.92	4.59	2.35
	Empty	3.00	2.40	1.97	1.63	1.08
12	Filled	10.98	8.70	6.80	5.18	2.59
	Empty	3.29	2.62	2.14	1.81	1.16

*Adapted from Ref. 54. R values (defined in text) do *not* include the sums of the effect of air film or surface conductance on the inside of the walls ($1/f_i = 0.68$) and on the outside ($1/f_o = 0.17$).
**Loose-fill insulation such as perlite, vermiculite, or others of similar density.

Table 3-10. Heat Transmission Values of Some Typical Concrete Masonry Walls*

Components	Heat resistance values, R**							
	A	B	C	D1	D2	E	F1	F2
Surface film (outside)	0.17	0.17	0.17	0.17	0.17	0.17	0.17	0.17
8-in. hollow concrete masonry at 100-pcf density (cores open)	1.75			1.75	1.75			
Cores filled with bulk insulation			4.85				4.85	4.85
Concrete brick at 140-pcf density		0.44				0.44		
2-in. air cavity		0.97						
4-in. hollow concrete masonry at 100-pcf density		1.40				1.40		
2-in. bulk insulation in cavity						4.00		
Batt insulation between furring strips: 1 in.				3.70			3.70	
2 or 2¼ in.					7.00			7.00
½-in. gypsum board interior finish				0.45	0.45		0.45	0.45
⅝-in. plaster, lightweight aggregate		0.39				0.39		
Surface film (inside)	0.68	0.68	0.68	0.68	0.68	0.68	0.68	0.68
TOTAL resistance, R	2.60	4.05	5.70	6.75	10.05	7.08	9.85	13.15
U value, 1/R	0.384	0.247	0.175	0.148	0.100	0.141	0.102	0.076

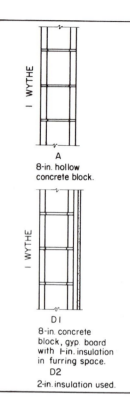

A
8-in. hollow concrete block.

D1
8-in. concrete block, gyp. board with 1-in. insulation in furring space.
D2
2-in. insulation used.

*Adapted from Ref. 54. The overall heat transmission coefficients, called U values, are determined by the heat resistance values, R, as explained in the text.
**Key for wall design shown at right.

other common types of wall components. In the latter table the U value for each complete assembly (bottom line) was computed after the R value of each component was tabulated and then added to give the total R value of each wall. In Table 3-11 the U values of 8-in. walls with various combinations of finish and insulation are given directly.*

It is important to select the correct wall as insulation against the weather, but this should not obscure some larger considerations affecting the total cost of heating and cooling a building. Heat flow through a single pane of glass in winter is 1.13 Btu per square foot per degree Fahrenheit per hour. This is six or seven times the heat flow through a lightweight concrete block wall having filled cores.

In general, the heat flow through the walls of a building is only a small part of the total heat flow. Heat losses through roofs, floors, and fenestration are other important considerations, as are losses due to infiltration, exhausts, fresh air intakes, and opening of doors.**

Effect of Moisture Content

Heat transfer (conductivity) values increase in walls as the moisture content increases. The relationships found† between the conductivity and the density and moisture content of concrete, mortar, or grout are shown in Fig. 3-17. The heat transfer values for oven-dry weights of concrete are frequently used. However, in an occupied building, the amount of moisture in the concrete or block may be termed "normally dry." This condition describes the equilibrium moisture content of concrete after extended exposure to 78 deg. F. air at 35 to 50% relative humidity. Note in Fig. 3-17 that conductivity in the normally dry condition is only slightly greater than in the oven-dry condition.

In the event that concrete masonry should become saturated with water, heat transfer will increase markedly. This condition could occur in a basement wall that is not dampproofed. In walls above grade the condition might prevail for a short time after a heavy driving rain.

A situation to avoid is having a sealer applied to the cold side of a wall. Moisture travels towards that side and, if its surface is sealed, the wall can approach

*See Ref. 49 for insulating values of other lightweight-aggregate concrete masonry walls.
**See Ref. 32.
†See Ref. 30.

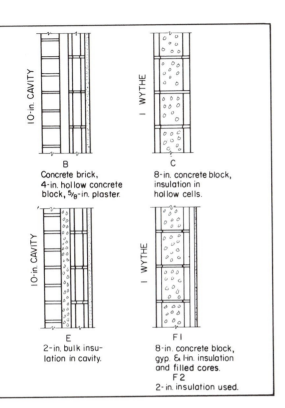

B
Concrete brick,
4-in. hollow concrete
block, ⅝-in. plaster.

C
8-in. concrete block,
insulation in
hollow cells.

E
2-in. bulk insu-
lation in cavity.

F1
8-in. concrete block,
gyp. & 1-in. insulation
and filled cores.
F2
2-in. insulation used.

Table 3-11. Estimated *U* Values for 8-In. Hollow Concrete Masonry Walls*

Wall details	U value based on density of concrete used in block				
	60 pcf	80 pcf	100 pcf	120 pcf	140 pcf
No insulation	0.32	0.34	0.38	0.43	0.55
No insulation, ½-in gypsum board on furring strips	0.21	0.23	0.25	0.27	0.31
No insulation, ½-in foil-backed gypsum board on furring strips	0.15	0.15	0.16	0.17	0.19
Loose-fill insulation in cores	0.12	0.14	0.18	0.21	0.35
Loose fill in cores, ½-in. gypsum board on furring	0.10	0.12	0.14	0.17	0.24
Loose fill in cores, ½-in. foil-backed gypsum board, furring	0.08	0.10	0.11	0.12	0.16
1-in. rigid glass fiber, ½-in gypsum board	0.14	0.14	0.15	0.15	0.17
1-in. polystyrene, ½-in. gypsum board	0.12	0.12	0.12	0.13	0.14
1-in. polyurethane, ½-in. gypsum board	0.10	0.10	0.11	0.11	0.12
Loose fill in cores plus 1-in. rigid glass, ½-in. gypsum board	0.08	0.09	0.10	0.11	0.14
Loose fill in cores plus 1-in. polystyrene, ½-in. gypsum board	0.07	0.08	0.09	0.10	0.12
Loose fill in cores plus 1-in. polyurethane, ½-in. gypsum board	0.07	0.07	0.08	0.09	0.11
R-7 blanket insulation, ½-in. gypsum board, furring	0.09	0.10	0.10	0.10	0.11

*Adapted from Ref. 55. *U* values of other wall types and sizes are also given in Ref. 55.

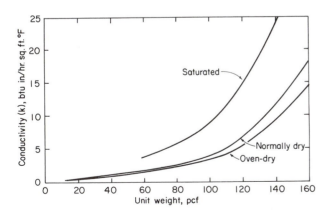

Fig. 3-17. Conductivity of concrete, mortar, or grout as affected by density and moisture. Aggregate type influences conductivity only as it affects the resulting unit weight and moisture content.

saturation there, thus losing much of its insulating value. This illustrates the importance of using exterior paints that can "breathe," as discussed in Chapter 7.

Heat Gain

The same heat transfer coefficients are used in both heating and cooling design considerations, except that slight differences are made in the heat losses of the wall surfaces themselves. In addition, the calculation of cooling load may make use of temperature differentials that account for the effects of solar radiation and exterior surface reflectances as well as the mass or weight of the building parts.

Heat gain from solar radiation is affected by the orientation of the building and the reflectivity of the exterior surfaces. Light-colored surfaces will reflect the sun's rays much more effectively than dark-colored surfaces. This explains the wider use of white or light-shaded paints and finishes for wall surfaces in the southern regions of the United States.

Steady State Versus Dynamic Thermal Response

A distinction must be made between masonry and nonmasonry construction because of the difference in heat storage capacity. Heavy construction such as concrete masonry does not respond to temperature fluctuations as rapidly as does lightweight construction, even though the U values may be identical. Due to a "fly-wheel effect" the net transfer of heat through a concrete masonry wall section for a certain time period might actually be less. This cyclic phenomenon is termed "dynamic thermal response."

The practice in calculating heat transfer as outlined previously is based on a "steady-state" temperature differential or a constant difference between extreme outdoor and indoor air temperatures. The size of heating or cooling equipment is calculated on this basis, but higher U values are permitted in buildings with heavy walls, floors, and roofs because of the long-known fact that they act as heat reservoirs. This is of importance since the actual temperature differential between indoors and outdoors is not constant but fluctuates with the time of day.

The steady-state technique has been the accepted method of design for many years, although the theory and basic mathematics of cyclic temperature heat losses have been known for more than a century and a half. The complexity and expense of the calculations prevented engineers from using the sophisticated approach until the era of the electronic computer arrived.

To analyze the effect of daily temperature changes, a computer program for dynamic thermal response was developed by the National Institute of Standards and Technology (previously called the National Bureau of Standards).* The computer program was verified with a structure built of 8-in. solid lightweight concrete masonry walls and a 4-in.-thick concrete roof. After the experimental structure was subjected to an outside cyclic temperature variation of 60 deg. F., it was confirmed that the heat flow rates calculated by the steady-state method ranged from 32 to 69% higher than the rates of the dynamic thermal response program.

The NBS computer program was used in a PCA study** to compare heat losses of walls having equal U values. Fig. 3-18 depicts the winter heat losses of three types of walls subject to typical outdoor temperature changes. The times at which peak loads occur are different, and there is a significant difference in the peak rate of heat loss, the value traditionally used to size heating equipment.

Fig. 3-19 shows summer heat gain curves for two types of walls with equal U value. It illustrates the well-

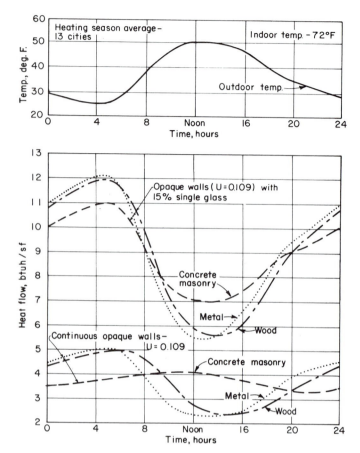

Fig. 3-18. Heat loss through masonry and nonmasonry walls, U values being equal.

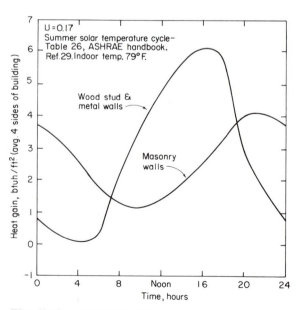

Fig. 3-19. Heat gain through masonry and nonmasonry walls, U values being equal.

*See Ref. 31.
**See Ref. 37.

known fact that during a hot day a masonry building is cooler than a lighter building. The relative air-conditioning loads can be visualized for the two types of walls.

Control of Water Vapor Condensation

When warm, humid air is chilled to a certain point, as by a cold surface, condensation takes place (Fig. 3-20). To prevent condensation or sweating on the interior or room-side surface of a wall, the overall resistance to heat transmission of the wall must be such that the surface temperature will not fall below the dew (condensation) point of the room air. Dew points for various room temperatures and relative humidities are listed in Table 3-12. It shows, for example, that with a room temperature of 70 deg. F. and relative humidity of 40%, the wall surface temperature should not fall below 45 deg. F. if sweating is to be avoided.

Normally, sweating of the interior surfaces of building walls is not a problem when proper attention is given to the insulating quality of the walls and the relative humidity is controlled within reasonable limits. Water vapor condensation that occurs on uninsulated basement walls below grade during humid periods of the summer or during clothes laundering activity may require mechanical dehumidification or ventilation for its control.

The control of condensation within wall spaces and wall materials is more complex because it can occur whether the wall has an overall high or low resistance to heat flow. Water vapor is a gas in the air and will diffuse through materials of building construction at rates that depend upon the water vapor permeability of the materials and the existing vapor pressure differential. The passage of water vapor through a material is not harmful in itself. It becomes of consequence only when, at some point along the water vapor flow path, a

Fig. 3-20. Beads of water on the outside of this tumbler of ice water are caused by condensation of water vapor in the air as it comes in contact with the cold surface.

temperature level drops below the dew point and condensation occurs.

Excessive accumulation of condensed water vapor in concrete masonry may lead to efflorescence or temporary formation of frost within the wall. In most concrete masonry buildings freezing does not constitute a problem due to the daily fluctuations in temperatures. Frost thaws and is released through the outer wythe as vapor or through weepholes as condensate.

The risk of damage from frost buildup in walls is greatest during extended periods of very low temperatures outdoors and relative humidities exceeding about 50% indoors. However, when outdoor temperatures

Table 3-12. Dew Point Temperatures

Dry bulb or room temperature, deg.F.	Dew point, deg.F., based on relative humidity									
	10%	20%	30%	40%	50%	60%	70%	80%	90%	100%
40	−9	5	13	19	24	28	31	34	37	40
45	−5	9	17	23	28	32	36	39	42	45
50	−1	13	21	27	32	37	41	44	47	50
55	3	17	25	31	37	41	45	49	52	55
60	6	20	29	36	41	46	50	54	57	60
65	10	24	33	40	46	51	55	58	62	65
70	13	28	37	45	51	56	60	63	67	70
75	17	31	42	49	55	60	65	68	72	75
80	20	36	46	54	60	65	69	73	77	80
85	23	40	50	58	65	70	74	78	82	85
90	27	44	55	62	69	74	79	82	86	90

are very low, high relative humidities indoors are rare except in laundries, bathing areas, etc. The maximum humidity that can be continuously maintained in a room is limited by the dew point of the wall or window or any sensitive zones on them.

To minimize condensation and yet provide for human health and comfort, some agencies* recommend that the maximum relative humidity maintained indoors should be approximately as follows:

Outdoor temperature, deg. F.	Indoor relative humdtiy, percent
0	25
10	30
20	35
30	40

An example of vapor pressure gradients appears in the lower part of Fig. 3-16 for a steady-state condition in which moisture and heat are migrating to the outdoors. The vapor pressure is expressed in inches of mercury (Hg). Gradient P_s assumes saturated air (100% relative humidity) indoors, and gradient P_c assumes continuous vapor flow for the actual humidities but neglects the possibility of condensation. Condensation would occur at the point where gradients P_s and P_c intersect. Since that point is in the granular vermiculite fill where resistance to vapor flow is very low, the condensation would probably take place on the inside of the outer wythe. Then, since the temperature there is about 5 deg. F., frost would develop. Gradient P_a shows actual vapor pressure gradient.**

Generally when the designer is confronted with building service conditions conducive to water vapor condensation, a vapor barrier on or as close as possible to the warm surface of the wall is provided. The vapor barrier is a material—such as plastic film, asphalt-treated paper, and aluminum or copper foil—that will transmit not more than one grain of water vapor per square foot of surface (normal to the vapor flow path) per hour under a vapor pressure differential of 1 in. of mercury. Thus, the vapor barrier will reduce to a minimum the entrance of water vapor into the wall. Then the reduced amounts of water vapor penetrating the vapor barrier will pass through the outer layers of the wall by diffusion. Of course, any outer surface treatment such as paint must be of a type that will "breathe."

For a vapor barrier to be fully effective, it must be applied as a leakproof, continuous layer. Openings such as those provided by electrical outlet boxes must be given close attention, and avenues for vapor leakage through or around window and door frames must be stopped.

Acoustics

Noise is unwanted sound, and what is considered noise depends upon the individual and the level of tolerance.

People value their privacy and do not care to hear the movements of their neighbors.

Acoustics as a science and technology is well advanced. The subject is complex, but if a few simple concepts are learned, the architect, engineer, and builder can do a great deal to assess and solve acoustical problems. Good acoustical design in residences and offices can be achieved with absorptive surfaces and relatively heavy wall and floor construction. Concrete masonry is an excellent sound barrier because of its density. An extended discussion of acoustics in buildings is presented in PCA's *Acoustics of Concrete in Buildings,* IS159T.†

Elements of Sound

The two important parameters used for the study of acoustics are frequency and decibels—or, loosely speaking, tone and loudness. The tone of a sound depends on the number of vibrations per second, the frequency. One vibration per second or one cycle per second (1cps) is called a hertz, abbreviated Hz. On the piano, middle C has a frequency of 262 Hz. The lowest of the 88 piano keys has a frequency of 27 Hz and the highest a frequency of 4186 Hz. This is essentially the range of tones used in the study of building acoustics.

Pressure of sound is measured by the decibel (db). Each increase of 20 db indicates a tenfold pressure increase. However, the ear mechanism automatically reduces its sensitivity as the pressure increases. An increase of 10 db is a threefold increase in pressure but the loudness sensation to the ear is only doubled. Since the human ear has a greater range of sensitivity to intensity in the middle range of frequencies, loudness is determined by the pressure level, decibels, at a frequency of 1,000 cps (1000 Hz).

An idea of the decibel as a unit of measure of sound intensity is obtained from the following list:

Decibels	Sound
140	Jet at takeoff 80 ft. from tail
130	Threshold of discomfort
120	Thunder
110	Car horn at 3 ft.
100	Wood saw at 3 ft.
90	Noise inside a city bus
80	Noisy office or vacuum cleaner
70	Boeing 707 jet landing (at 3,300 ft.)
60	Average office
50	Average conversation
40	Quiet radio

*See Ref. 41.
**See Refs. 29 and 32.
†See Ref. 28

30 .... Quiet conversation
20 .... Whisper at 4 ft.
10 .... Normal breathing
0 .... Threshold of audibility

Criteria for Acoustic Ratings

Building codes regulate the amount of noise that must be stopped by walls, floors, and ceilings. Typical sound loss values are in the range of 40 to 55 db for airborne and impact sounds.

There are three principal types of ratings, as shown in Table 3-13. Each type may be identified at any individual frequency or by class. Sound absorption coefficient (SAC) and noise reduction coefficient (NRC) values are in sabins, which are units measuring energy absorption and not sound or loudness. All other ratings are in decibels. The ratings have one thing in common: *the larger the number, the better the sound insulating quality of the wall or floor.*

"Sound absorption" refers to the amount of airborne sound energy (sabins) absorbed by the wall surface adjacent to the sound. "Sound transmission loss" is the total amount of airborne sound (decibels) lost as it travels from one side of a wall or floor to the other. "Impact noise isolation" and "impact sound isolation" refer to the number of decibels lost through a floor from standardized impacts on top of the floor (or floor covering). ISPL, INR, and IIC ratings are not used for walls. The ratings are governed by ASTM C423, C634, E90, E336, E413, E492, and E597.

Sound Absorption

Fig. 3-21 shows data for 8-in. lightweight concrete masonry walls tested in two different laboratories. Note how the absorption varies with the frequency. The sound absorption coefficient (SAC) is the amount of sound energy absorbed compared to a perfectly ab-

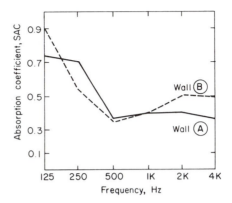

Fig. 3-21. Sound absorption test data for two 8-in. concrete masonry walls made with shale aggregate.

sorptive surface such as an open window. Translated into decibels, SAC at 0.9 equals only 10 db; at 0.7, 5 db; at 0.5, 3 db; and at 0.3, 1.5 db. This is true regardless of the sound level. As a practical matter, it is difficult to lose much more than about 5 db of sound by absorption (reduce the loudness by 33%).

The noise reduction coefficient (NRC) is found by averaging the SAC values at frequencies of 250, 500, 1000, and 2000 Hz. In Fig. 3-22, Wall A has an NRC value of 0.46 (about 3 db).

Typical NRC values are given in Table 3-14. However, a concrete block with uniformly fine texture, as shown at the top of Fig. 1-32, may have an NRC value as high as 0.68 (5 db). In contrast, glass, plaster, and other smooth surfaces have NRC values of less than 0.05 (less than one-sixth of a decibel). Also note in Table 3-14 that *painted* concrete masonry has a reduced NRC value.

Table 3-13. Acoustic Ratings

Type of rating	Rating by individual frequencies	Rating by class
Sound absorption (airborne)	SAC	NRC
Sound transmission loss (airborne)	STL	STC
Impact noise isolation or Impact sound isolation	ISPL	INR
		IIC

Table 3-14. Approximate Noise Reduction Coefficients*

Material	Surface texture	Approximate NRC
Lightweight aggregate block, unpainted	Coarse	0.40
	Medium	0.45
	Fine	0.50
Heavy aggregate block, unpainted	Coarse	0.26
	Medium	0.27
	Fine	0.28
Deduct from above for painting		

Paint	Application	For 1 coat	For 2 coats
All	Spray	10%	20%
Oil	Brush	20%	55%
Latex	Brush	30%	55%
Cement	Brush	60%	90%

*Adapted from Ref. 100.

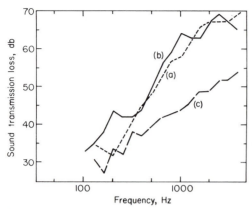

Notes:

Wall (a) is 8-in. hollow heavyweight concrete masonry, one side painted and the other side furred, plus 1/2-in. gypsum wallboard. Weight is 45.6 psf and STC is 49.

Wall (b) is 6-in. hollow lightweight concrete masonry, one side painted and the other side 1/2-in. gypsum board on resilient channel furring. Weight is approximately 26 psf and STC is 53.

Wall (c) is same as (b) but without finish. Weight is about 23 psf and STC is 44.

Fig. 3-22. Sound transmission loss test date for several concrete masonry walls.

Sound Transmission Loss

If walls have sealed surface pores on at least one side, the sound transmission loss (STL) is related to the weight of the wall. When the pores are sealed, the wall tends to transmit sound by acting as a diaphragm, literally stopping sound from passing through it. Because of this, a heavier wall tends to reduce the transmission of sound more than a light wall, following what is known as "mass law." It holds true for homogeneous partitions that are nonporous and have uniform physical properties throughout the entire wall panel.

Table 3-15 shows the STC values for various types of concrete masonry walls. An example of STL test data for several concrete masonry walls is given in Fig. 3-22. Additional STC values up to 59 and 63 were obtained for concrete masonry and cast concrete walls, respectively, and the results are given in References 28 and 34.

All of the stated values for sound transmission loss through a wall are meaningless if the wall panel has an opening. Large (as a window opening) or small (as an opening for a water pipe or electrical conduit), an opening can seriously alter the sound reduction capability of a wall. If maximum sound reduction is desired, any opening should be carefully avoided. For example, placing electrical outlet boxes back to back through a concrete masonry wall will provide an opening or flanking path through which sound can travel.

Just as undesirable are any openings around doors, pipes, and air ducts. Another flanking path is provided by a partition that extends only to a suspended ceiling rather than to the floor or roof above. Calking is necessary where walls join and between walls and ceilings.

The sound level in a room is known as the masking sound or background noise; it fairly well swallows up lesser sounds transmitted through walls and floors. This is because the decibel level of separate sounds are not directly additive. Table 3-16 shows how sound transmission loss through a wall panel affects noise conditions. The values are based on an assumed background noise level that corresponds to average conditions in most residences. Background noise levels in almost all offices and most institutions are higher.

Impact Sound Isolation

Impact sound ratings were adopted to combat annoying sounds of footsteps and objects dropping above. For an effective sound barrier, a floor should have an impact sound isolation (IIC) rating of at least 40 db.

Although test data for floors made with concrete masonry are meager, other types of concrete floors have been tested many times. For example, here are the IIC ratings of several bare concrete floors:

	Decibels
4-in.-thick solid slab, 53 psf	25
6-in.-thick hollow prestressed slab, 43 psf	23
8-in.-thick hollow slab, 45-60 psf	26
10-in.-thick solid slab, 121 psf	29

It can be expected that concrete masonry will perform slightly better than concrete. In either case the sound isolation of the bare floor is marginal, but floor coverings and ceiling treatments are immensely helpful, as shown in Table 3-17. Floor or ceiling treatments that add at least 20 db to the ICC rating will solve the acoustical problem of impact noises on concrete floors.

Other Acoustical Considerations

Noise-producing equipment should be kept as far as possible from occupied areas, especially bedrooms. Flexible connectors should be used to couple mechanical equipment to pipes, ducts, and electric power. Also, pipes and ducts should not be firmly connected to parts of the building that can serve as sounding boards. Instead they should be supported by resilient connections to solid supports. Where they pass through walls and floors, they should be isolated from the construction by gaskets.

If a sound source must be close to a work area, sound barriers and sound absorbers around the source should be considered. In many instances quieter appliances and other equipment are the best solution. Manufac-

Table 3-15. Data from Sound Transmission Loss Tests (ASTM E90) of Concrete Masonry Walls*

Wall description	Test No.**	Wall weight, psf	STC
Unpainted walls:			
8-in. hollow lightweight-aggregate units, fully grouted, No. 5 vertical bars at approx. 40 in. oc	1023-1-71	73	48
8-in.. hollow lightweight-aggregate units	1144-2-71	43	49
8-in. composite wall—4-in. brick, 4-in. lightweight hollow units	1023-4-71	58	51
8-in. dense-aggregate hollow units	1144-3-71	53	52
10-in. cavity wall—4-in. brick, 4-in. lightweight hollow units	1023-6-71	56	54
Walls painted on both sides with 2 coats of latex paint:			
4-in. hollow lightweight-aggregate units	1379-5-72	22	43
4-in. hollow dense-aggregate units	1379-3-72	29	44
6-in. hollow lightweight-aggregate units	933-2-70	28	46
6-in. hollow dense-aggregate units	1379-1-72	39	48
8-in. hollow lightweight-aggregate units, fully grouted, No. 5 vertical bars at approx. 40 in. oc	1023-2-71	73	55
Walls plastered with ½-in. gypsum plaster on both sides:†			
8-in. composite wall—4-in. brick, 4-in. lightweight hollow units	1023-10-71	61	53
8-in. hollow lightweight-aggregate units, fully grouted, No. 5 vertical bars at approx. 40 in. oc	1023-9-71	79	56
10-in. cavity wall—4-in. brick, 4-in. lightweight hollow units	1023-8-71	59	57
Walls covered with ½-in. gypsum board on resilient channels:†			
4-in. hollow lightweight-aggregate units	1379-4-72	26	47
4-in. hollow dense-aggregate units	1379-2-72	32	48
8-in. composite wall—4-in. brick, 4-in. lightweight hollow units	1023-5-71	60	56
8-in. hollow lightweight-aggregate units	933-1-70	40	56
10-in. cavity wall—4-in. brick, 4-in. lightweight hollow units	1023-7-71	58	59
8-in. hollow lightweight-aggregate units, fully grouted, No. 5 vertical bars at approx. 40 in. oc	1023-3-71	77	60

*Adapted from Table 2, Ref. 23.
**Kodaras Acoustical Laboratories, Elmhurst, N.Y.
†Surface treatment on block side only of composite and cavity walls.

turers have become aware of the need and many "sound-rated" devices are now available, frequently at little additional cost.

When a single wall is used as a sound barrier, it is sometimes desirable to provide a resilient connection between the wall and the building frame. A 5- to 7-db improvement can result. Double walls perform better than single walls when of equal weight. Increased separation and sound-absorbing material in the cavity add to the performance of cavity walls. The sound transmission loss of a cavity wall is frequently about 8 db better than a solid wall of equal weight and, if the two wythes are of unequal weight, as much as 4 db more can be added to the STC rating.

Table 3-16. Relationship Between Sound Transmission Loss Through a Wall and Hearing Conditions on Quiet Side*

Transmission loss, db	Hearing condition	Rating
30 or less	Normal speech can be understood quite easily and distinctly through the wall.	Poor
30 to 35	Loud speech can be understood fairly well. Normal speech can be heard but not easily understood.	Fair
35 to 40	Loud speech can be heard but is not easily intelligible. Normal speech can be heard only faintly, if at all.	Good
40 to 45	Loud speech can be faintly heard but not understood. Normal speech is inaudible.	Very good—recommended for dividing walls between apartments.
45 or more	Very loud sounds such as loud singing, brass musical instruments, or a radio at full volume can be heard only faintly or not at all.	Excellent—recommended for band rooms, music practice rooms, radio and sound studios.

*This table is based on the assumption that a noise corresponding to 30 db is continuously present on the listening side.

Table 3-17. Acoustic Rating Improvements for Concrete Floors with Various Treatments

Treatment	IIC (impact) improvement, db
Carpet, pad, and acoustic ceiling	58
Carpet and pad	48
Acoustic ceiling	27
½-in. T&G wood parquet	25
Cork tile and furred ceiling	21
Plaster or gypsum board ceiling	8
Vinyl tile	4
2-in. concrete topping	0

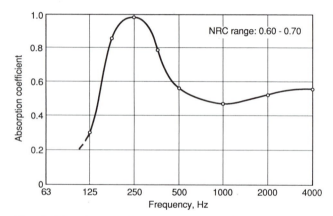

Fig. 3-23. Sound absorption test data for concrete "sound block". Also see Fig. 1-29a.

Selective absorption that matches the frequency of an unwanted sound is obtained by slotting the face shells of concrete masonry units, Fig. 1-29a and Fig. 3-23. Fig. 3-24 here shows the result: an attractive and sound-absorbing wall.

Highway and Railway Sound Barriers

Those who live or work along heavy traffic routes are well aware of traffic noise, typified in Table 3-18. Concrete masonry walls (Fig. 3-25) are helpful in reducing this noise level and at the same time provide a measure of privacy.

Studies* have shown that a 15-db sound reduction can readily be achieved by a concrete masonry wall. The level of sound reduction depends on the relative distances of the wall to the highway (or rail line) and

to the listener as well as the height of the wall above the line of sight to the vehicle.

Fire Resistance

Firesafety is a major element of building codes. Fire-related code provisions address both life-safety and property protection concerns by requiring building components and assemblies to meet a specified level of fire resistance. In simplistic terms, fire resistance can be defined as that property of a material or an assembly to withstand fire, or to give protection from it (Fig. 3-26).

*See Ref. 35.

Fig. 3-24. Slotted concrete block improve the acoustics.

Table 3-18. Typical Noise Levels of Transportation Vehicles

Type of traffic	Noise level, db, based on distance from vehicle		
	50 ft.	200 ft.	800 ft.
Passenger car, 50-60 mph	70	58	46
Truck, max. highway speed	88	76	64
Diesel train, 30-50 mph	97	85	73

Fig. 3-25. Traffic noise from expressways can be effectively shielded by a concrete masonry wall.

As building components, walls can have a dual function in their performance against fire. When they are constructed solely to prevent the spread of fire to another compartment or building, they are said to possess barrier fire resistance. This type of fire resistance would pertain to fire-rated, non-load-bearing walls. When walls are only required to perform structurally (exterior load-bearing walls sufficiently distant from a property line) for a specified fire rating period, they are said to possess structural fire resistance. It naturally follows that there are situations where walls must possess both barrier and structural fire resistance. Typically, the structural fire resistance of a concrete masonry wall will exceed the fire resistance that the wall can achieve as a barrier. Therefore, a concrete masonry wall will normally continue to carry a load, even though its established fire resistance rating period has been exceeded.

Whether walls are required to perform structurally, or as a barrier, or both, concrete masonry makes an excellent choice of building material. With its design flexibility, architects can adapt concrete masonry to a variety of prescribed fire resistance requirements for walls. Filling of core spaces, selecting different aggregates to change the unit weight, as well as utilizing different block sizes are all ways of accomplishing this. Concrete masonry walls also display excellent durability, such that only one test specimen is needed when

Fig. 3-26. Fire resistance is but one of the many positive qualities of concrete masonry, giving a feeling of safety.

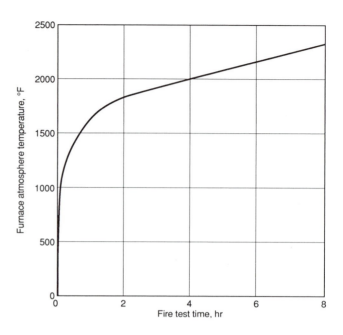

Fig. 3-27. ASTM Standard E119 time-temperature curve for fire tests.

conducting the ASTM E119 standard fire test. This is explained in greater detail later. Still another good quality of concrete masonry is it doesn't burn, and therefore gives off no smoke or toxic fumes during a fire.

How Fire Endurance is Determined

Fire endurance can be defined as a measure of the elapsed time during which a material or assembly continues to exhibit fire resistance under specified conditions of test and performance. Fire endurance periods of building components are typically based on physical tests conducted in accordance with ASTM E119, Standard Methods of Fire Tests of Building Construction and Materials. Provisions of the ASTM E119 require that test specimens be subjected to a standard fire condition that follows the time-temperature relationship shown in Fig. 3-27.

Under the E119 standard, the fire endurance of a wall assembly is determined by the time required to reach the first of any of the following test end points. Items 1 and 2 are associated with the wall's ability to perform as a barrier, whereas the third item provides a criterion for structural evaluation.

1. The ignition of cotton waste due to the passage of flame or hot gases through cracks or fissures.
2. A temperature rise of 325 deg. F (single point) or 250 deg. F (average) above ambient on the unexposed surface of the wall. This is known as the heat-transmission end point.
3. An inability to carry the applied design load, or structural collapse of the wall. This is known as the structural end point.

Upon completion of the fire resistance portion of the fire endurance test (Fig. 3-28), the wall assembly must also be subjected to the impact, erosion, and cooling effects of a standard hose stream test (Fig. 3-29). The duration of the water application on the fire-exposed side of the wall is based on the duration of the specimen's fire rating period. Any projection of water that passes through the wall's unexposed surface during the test constitutes an unsuccessful performance and terminates the test.

Fig. 3-28. A concrete masonry wall undergoing a fire test. A gas-fired furnace is on the opposite side of the wall. Jacks under the wall impose a live load on the wall during the test, and every phase of the test is closely observed and recorded by engineers.

Fig. 3-29. Fire hose stream test being applied at incandescent face of concrete masonry wall immediately after it is removed from the furnace.

It should be noted that the hose stream test is not intended to duplicate the scenario of water from a fire-fighters hose striking a wall in an actual fire. Rather, it is intended to evaluate the test specimen's resistance to thermal shock. By design, the purpose of the hose stream test is to examine the specimen's ability as a fire barrier, and to take a conservative approach in guarding against the structural collapse of a wall under real fire conditions.

Table 3-19 summarizes the ASTM E119 end-point criteria and test conditions for wall assemblies.

Introduction to Analytical Methods for Calculating Fire Endurance

The three U.S. model building codes (BOCA/NBC, SBC, and UBC)* have traditionally required standardized fire testing to satisy fire-resistance rating requirements for walls and other building components. Today, however, analytical calculation methods are recog-

nized and accepted in the codes for the determination of fire endurance. This has been made possible due to the wealth of test data that has been compiled from ASTM E119 tests through the years. The advantage of using analytical methods is the significant cost savings realized compared to the practice of conducting full-scale fire tests.

To analytically calculate the fire endurance of a wall, one must understand the end-point criteria of the standard testing procedure that governs its fire rating. The terminating point, or end-point condition, of an ASTM E119 test can occur as any one of the four criteria described previously (including the hose stream test). When concrete masonry walls are tested, the heat transmission end-point is nearly always reached prior to the passage of flame or gases, or structural failure. Thus, heat transmission is the controlling factor in establishing the fire resistance rating period assigned to the wall.

See Ref. 79, 80, and 81.

Table 3-19. Applicable End-Point Criteria and Test Conditions for Concrete Masonry Members and Assemblies (Based on ASTM E119 Standard Fire Tests)

Member		End point			
		250°F average temperature rise or 325°F point temp. rise on unexposed surface	Flame impingement through cracks or fissures sufficient to ignite cotton waste	Carry applied load	Hose stream test
Walls	Bearing	Yes	Yes	Yes	Yes*
	Nonbearing	Yes	Yes	No load applied	Yes*

*Hose stream tests apply only to those walls required to have a one-hour rating or greater.

Concrete masonry also performs well in the hose stream portion of the test because of its exceptional durability. This durability permits the same test specimen that is subjected to the burn portion of the test for the full fire endurance period to also be used in the hose stream test immediately afterward. Other assemblies, such as those utilizing wood or steel studs and gypsum wallboard, are routinely tested by subjecting a duplicate test specimen to the hose stream test after being fire tested for only one-half of the fire endurance period of the original test specimen. This doesn't seem equitable, but nevertheless, it is permitted under the ASTM E119 test standard.

Factors Influencing Fire Endurance (Heat Transmission)

For concrete masonry units, equivalent thickness of the block (defined below) and unit weight are the primary factors that influence fire endurance. Moisture content does affect the material's thermal conductivity, but this is described in another section.

As indicated above, the fire endurance of concrete masonry walls will most often be determined by the heat-transmission end point. Since this parameter is a function of the temperature distribution through the wall, the fire endurance can be predicted analytically if the type of aggregate and equivalent thickness of the masonry units are known.

Effect of Aggregate on Unit Weight and Fire Endurance

As unit weight, which is determined by aggregate type, is reduced, the resistance to heat transmission increases, thereby increasing the fire endurance. Examination of Table 3-20 confirms that a concrete masonry wall utilizing lighter-weight aggregate units provides greater fire endurance than one constructed of normal-weight aggregate units. This is reflected by the requirement of

a lesser thickness of lighter-weight aggregate block to achieve the same fire endurance rating as a normal-weight aggregate unit.

Structural lightweight concretes use aggregates such as expanded shale, clay, and slate and have unit weights ranging from 100 pcf to about 120 pcf. Normal-weight concretes have unit weights ranging from 135 to 150 pcf. Normal-weight concretes utilize siliceous aggregates obtianed from natural sand and gravel or carbonate aggregates such as limestone.

Equivalent Thickness

Referring again to Table 3-20, the other critical parameter affecting the fire endurance of concrete masonry walls is the equivalent thickness. This factor is determined from the wall's actual thickness and the geometry of the concrete masonry unit. Hollow or solid concrete masonry units are available in nominal thicknesses of 4, 6, 8, 10, and 12 in. with varying percentages of solid area. The equivalent thickness of a block can be calculated using the following equation:

$$T_{eq} = \% \text{ solid} \times \text{actual thickness} \qquad \text{(Eq. 1)}$$

For example, the equivalent thickness of a 50% solid, 8-in. nominal block would be calculated as: $0.50 \times 7.625 = 3.8$ in. The percent of solids of any given masonry unit can be obtained from the manufacturer. If 100% solid flat-sided masonry units are used, the equivalent thickness is the actual thickness.

Blended Aggregate Block

When it is necessary to determine the equivalent thickness of blended aggregate block, the *Uniform Building Code* (UBC) can be referenced. It contains a method that is based on the interpolation of Table 3-20 values in proportion to the percentage of the volume occupied by each of the block's component aggregates. Under the provisions of the UBC, the method is applicable to blends composed of pumice, or expanded aggregates

Table 3-20. Minimum Equivalent Thickness in Inches of Load-Bearing Concrete Masonry Unit Walls for Fire-Resistance Ratings*

A. SBC**	4 hr.	3 hr.	2 hr.	1 hr.
Pumice or expanded slag aggregates	4.7	4.0	3.2	2.1
Expanded shale, clay, or slate aggregates	5.1	4.4	3.6	2.6
Limestone, cinders, or unexpanded slag aggregates	5.9	5.0	4.0	2.7
Calcareous gravel aggregates	6.2	5.3	4.2	2.8
Siliceous gravel aggregates	6.7	5.7	4.5	3.0
B. UBC and BOCA/NBC†**	**4 hr.**	**3 hr.**	**2 hr.**	**1 hr.**
Pumice or expanded slag	4.7	4.0	3.2	2.1
Expanded clay, shale, or slate	5.1	4.4	3.6	2.6
Limestone, cinders, or air-cooled slag	5.9	5.0	4.0	2.7
Calcareous or siliceous gravel	6.2	5.3	4.2	2.8

*SBC—Fire ratings for thicknesses between tabulated values may be obtained by direct interpolation. UBC and BOCA/NBC—For thicknesses between tabulated values, refer to the expanded tables in the codes.

**SBC and UBC—Where all of the core spaces of hollow-core wall panels are filled with loose-fill material, such as expanded shale, clay or slag, or vermiculite or perlite, the fire-resistance rating of the wall is the same as that of a solid wall of the same concrete type and of the same overall thickness.

†In the BOCA/NBC, 2-hour fire-resistance rated walls composed of hollow concrete masonry units having a nominal thickness of 8 in. or greater are permitted to be classified as having 4 hours of fire resistance when all of the core spaces of the units are filled. Grout, insulation, or dry granular materials as described in the second note above are considered as acceptable fill.

such as slag, clay, shale, and slate, in combination with calcareous and siliceous gravel provided the gravel will pass through a No. 4 sieve. Note that the blending of limestone, cinders, or slag, with calcareous or siliceous aggregate concrete masonry is not addressed. The required equivalent thickness of a blended concrete masonry unit (wall) is determined as follows (aggregate reference numbers correspond to aggregate categories in Part B of Table 3-20).

For calcareous or siliceous gravel (aggregate 4) blended with pumice or expanded slag (aggregate 1):

$$T_{eq\ required} = (T_{eq\ agg\ 4} \times V_{agg\ 4}) + (T_{eq\ agg\ 1} \times V_{agg\ 1}) \quad \text{(Eq. 2)}$$

For calcareous or siliceous gravel (aggregate 4) blended with expanded clay, shale, or slate (aggregate 2):

$$T_{eq\ required} = (T_{eq\ agg\ 4} \times V_{agg\ 4}) + (T_{eq\ agg\ 2} \times V_{agg\ 2}) \quad \text{(Eq. 3)}$$

For calcareous or siliceous gravel (aggregate 4) blended with both categories of aggregates (aggregate 1 and 2):

$$T_{eq\ required} = (T_{eq\ agg\ 4} \times V_{agg\ 4}) + [T_{eq\ agg\ 2} \times (V_{agg\ 1} + V_{agg\ 2})] \quad \text{(Eq. 4)}$$

where,

$T_{eq\ agg\ 1}$, $T_{eq\ agg\ 2}$, and $T_{eq\ agg\ 4}$ = specified equivalent thickness values as given in Table 3-20.

$V_{agg\ 1}$, $V_{agg\ 2}$, and $V_{agg\ 4}$ = the ratio of the volume of individual categorized aggregate to the total volume of aggregate for aggregates indicated in Table 3-20.

The following example demonstrates the use of the formula.

Given: Concrete masonry units are composed of 15% pumice, 30% expanded shale, and 55% calcareous aggregate.

Determine the required equivalent thickness of the units needed to provide a 2-hour fire-rated wall.

From Table 3-20 and blending formula (Eq. 4),

$$T_{eq\ required} = 4.2\,(0.55) + 3.6\,(0.15 + 0.30)$$
$$= 3.9\ in.$$

The result indicates that the equivalent thickness of concrete masonry units composed of aggregates blended to the above proportions must be at least 3.9 in. if the wall is to provide a fire-resistance rating of 2 hours.

Effect of Fill

Completely filling the core spaces of hollow concrete masonry units with perlite, vermiculite, grout, or expanded slag, clay, shale, or slate, increases fire endurance such that the hollow units will perform as 100% solid units. This is due to the block's increased resistance to heat transmission provided by the fill material. Additional information on this subject is addressed in the notes to Table 3-20 as it applies to each of the model building codes. If grouting or filling of cores is only done intermittently along the length of the wall, the equivalent thickness—and thus the fire endurance of the wall—should be based on that of the hollow masonry units.

Multi-Wythe Walls

When multi-wythe walls are constructed of concrete masonry, the fire endurance of the composite wall is greater than the summation of the individual fire endurance periods of its component layers. An equation that can be used to estimate the fire endurance of multi-wythe walls based on the heat-transmission end point is:

$$R = (R_1^{0.59} + R_2^{0.59} + \ldots + R_n^{0.59})^{1.7} \qquad \text{(Eq. 5)}$$

where,

R = total fire endurance rating in minutes

$R_1, R_2, \ldots R_n$ = fire endurance in minutes of each individual wythe

When using the above equation, it is important to remember that this is just an estimation. Equation 5 is not applicable in all cases because the exponent 1.7 and its reciprocal 0.59 are average values that vary from material to material. Also, the equation does not account for the orientation of the wythes. For example, in actual testing, if the more fire-resistant material of

two wythes is in the outer layer exposed to the fire, a higher fire resistance rating will result than if the wythes were reversed and tested.

This is significant in code applications because building codes require that the fire ratings of walls of nonsymmetrical construction be based on the lesser of the two ratings obtained from tests conducted on each side of the wall. Exceptions to this are contained in the BOCA/NBC and SBC codes for exterior walls greater than 5 ft. from an interior lot line. When this situation is present, the fire rating is established using the interior side of the wall as the fire-exposed surface.

To apply the multi-wythe equation (Eq. 5) to composite concrete masonry walls, "R" values should be obtained from Table 3-20 and converted to minutes. Values for other materials, such as concrete and brick masonry, are published in other industry sources (Reference 82).

Finish Materials Applied to Concrete Masonry

The *Standard Building Code* (SBC) contains guidelines for estimating the additional fire endurance provided by plaster and gypsum wallboard finishes applied to the fire-exposed and/or non-fire-exposed sides of concrete masonry walls.

Where plaster or gypsum wallboard is applied to the non-fire-exposed side of the wall, the fire resistance contribution of the finish material is determined as follows: Based on the type of aggregate used in the concrete masonry, and the type and thickness of the finish material, a multiplying factor from Table 3-21 is applied to the actual thickness of the finish material to

Table 3-21. Multiplying Factor for Finishes on Non-Fire Exposed Side of Wall

Type of finish applied to wall	Type of aggregate used in concrete masonry			
	Siliceous or calcareous gravel	Limestone, cinders, or unexpanded slag	Expanded shale, clay, or slate	Pumice, or expanded slag
Portland cement-sand plaster	1.00	0.75*	0.75*	0.50*
Gypsum-sand plaster or gypsum wallboard	1.25	1.00	1.00	1.00
Gypsum-vermiculite or perlite plaster	1.75	1.50	1.25	1.25

*For portland cement-sand plaster ⅝ in. or less in thickness and applied directly to the concrete masonry on the non-fire-exposed side of the wall, the multiplying factor shall be 1.00.

Reproduced from the 1988 edition of the *Standard Building Code,* with the permission of the copyright holder, Southern Building Code Congress International, Inc. All rights reserved.

Table 3-22. Time Assigned to Finish Materials on Fire-Exposed Side of Wall

Finish description	Time, min.
Gypsum wallboard	
⅜ in.	10
½ in.	15
⅝ in.	30
2 layers of ⅜ in.	25
1 layer ⅜ in., 1 layer ½ in.	35
2 layers ½ in.	40
Type X gypsum wallboard	
½ in.	25
⅝ in.	40
Portland cement-sand plaster applied directly to concrete masonry	*
Portland cement-sand plaster on metal lath	
¾ in.	20
⅞ in.	25
1 in.	30
Gypsum-sand plaster on ⅜ in. gypsum lath	
½ in.	35
⅝ in.	40
¾ in.	50
Gypsum-sand plaster on metal lath	
¾ in.	50
⅞ in.	60
1 in.	80

*The actual thickness of portland cement-sand plaster, provided it is ⅝ in. or less in thickness, may be included in determining the equivalent thickness of the masonry for use in Table 3-20.

Reproduced from the 1988 edition of the *Standard Building Code,* with the permission of the copyright holder, Southern Building Code Congress International, Inc. All rights reserved.

obtain an adjusted thickness. This adjusted thickness is then added to the equivalent thickness of the concrete masonry to get an adjusted equivalent thickness, whereby Table 3-20 can be used to determine the fire resistance of the composite wall.

Where plaster or gypsum wallboard is applied to the fire-exposed side of the wall, the corresponding time assigned to the finish material shown in Table 3-22 is added directly to the fire resistance rating of the concrete masonry wall as determined from Table 3-20. Where finish materials are applied to both sides of the wall, the fire resistance contribution of the material on the non-fire-exposed side of the wall is determined first as described above. This adjusted value is then added directly to the appropriate Table 3-22 value to account for the fire resistance contribution of the material on the fire-exposed side of the wall. The net effect is the fire resistance of the entire assembly.

As is the case with nonsymmetrical multi-wythe wall assemblies, calculations to determine the fire resistance of walls having finish materials on one side, or

finishes of different types and thicknesses on each side, must be performed twice. In performing the calculations, the above procedure is used with the assumption that either side may be the fire-exposed side. The lesser of the two calculated values then establishes the fire resistance rating.

An exception applies in the SBC for exterior walls having a horizontal separation of more than 5 ft. from a lot line. In this case, the fire is assumed to occur on the interior side of the wall.

Protection of Structural Steel Columns

Another application of concrete masonry units is providing fire protection for structural steel columns. An empirical equation, based on a limiting condition of the column's steel core temperature reaching 1000 deg. F., is available for predicting the fire endurance of box-type concrete masonry-protected steel columns. The calculation method and equation are not provided in this text, but a comprehensive discussion of this subject, including examples, can be found in Reference 83.

Closing Comments

The intent of this section on fire resistance is only for background information purposes. While it is believed that the information contained herein accurately reflects the content of the current editions of the three U.S. model building codes, the reader is advised to consult the applicable building code and abide by its provisions if any discrepancies are noted.

Chapter 4

DESIGN AND LAYOUT OF CONCRETE MASONRY WALLS

The design of a concrete masonry wall depends on its required appearance, economy, strength, insulation, and acoustics. The layout of the wall involves other important considerations, such as the internal arrangement of components, modular planning, and provisions for shrinkage cracking control and weather resistance. All deserve careful planning if the wall is to successfully serve its intended purpose.

Types of Walls

Concrete masonry walls may be classified as solid, hollow, cavity, composite, veneered, reinforced, or grouted. These classifications sometimes overlap, but the basic terminology and bonding directions remain the same, as shown in Fig. 4-2.

Fig. 4-1. Customized concrete block arranged in a dominating pattern.

Solid Masonry Walls

Solid masonry walls (Fig. 4-3) are built of solid masonry units (ASTM C129 or C145) with all joints completely filled with mortar or grout. At least 75% of the net area of an ASTM C145 solid unit is solid concrete (cells are less than 25% of the unit). Facing units are usually brick or other solid architectural units that are laid with full head and bed joints. Backup units consist of solid masonry units laid with full head and bed joints.

If units with flanged ends are used, the end cavity must be filled with grout. Spaces between units are filled with mortar.

Structural bond between wythes is ensured by masonry headers, unit metal ties, continuous metal ties, or grout. Typical codes require that not less than 4% of the area of each wall face be composed of headers.

Headers usually consist of facing units laid transversely so they extend 3 to 4 in. into the backing. If the wall does not have headers extending completely through it, headers from the opposite side overlap 3 to 4 in. The allowable vertical or horizontal distance between adjacent headers varies from 24 to 36 in., depending on the local code. Fig. 4-3 shows a solid wall in which the courses of each wythe overlap to create headers.

Although common in past decades, solid masonry units are rarely used today. Due to this obsolescence, ASTM withdrew its C145 standard in 1990.

Hollow Masonry Walls

These walls (Fig. 4-4) are built of hollow (ASTM C90 or C129) or combined hollow and solid masonry units laid in face-shell mortar bedding.

Hollow masonry walls may be built in any required thickness with single or multiple wythes. Multi-wythe hollow masonry walls usually consist of two wythes—facing and backup—and may also be classified as composite walls. Bond between wythes is ensured by masonry headers, unit metal ties, continuous metal ties, or grout. All collar joints are filled with mortar.

Cavity Walls

A cavity wall (Fig. 4-5) is a multi-wythe wall with noncomposite action, allowing each wythe to independently accept and react to stress relative to its stiffness. A cavity wall consists of two walls separated by a continuous air space 2 in. or more wide and tied together by rigid metal ties embedded in the mortar joints of both walls. Cavities more than 4 in. wide must have ties designed to support the load without pullout or buckling to allow compatible lateral deflection between wythes. The facing wall usually consists of one wythe of solid or hollow masonry units 3½ to 4 in. thick. The backing may be a single- or multi-wythe solid or hollow masonry wall. The thickness of the

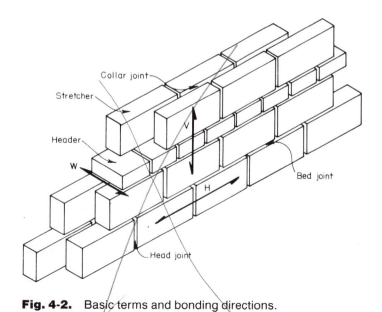

Fig. 4-2. Basic terms and bonding directions.

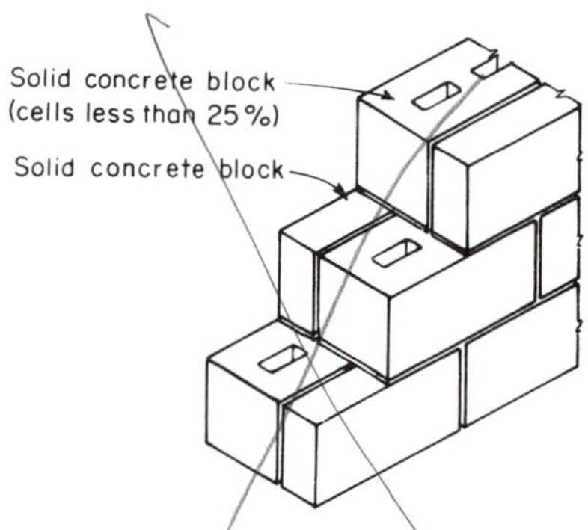

Collar joint

Stretcher

Header

W

H

Bed joint

Head joint

Solid concrete block (cells less than 25%)

Solid concrete block

Fig. 4-3. Solid masonry wall.

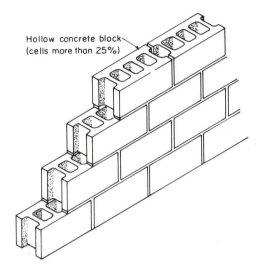

Fig. 4-4. Hollow masonry wall.

Hollow concrete block
(cells more than 25%)

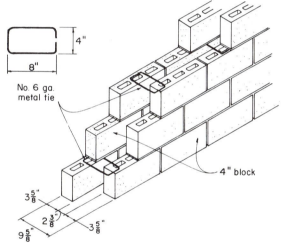

No. 6 ga. metal tie

4" block

(a) 10-in. wall of 4-in. block

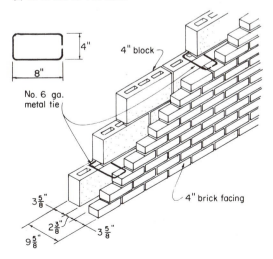

4" block

No. 6 ga. metal tie

4" brick facing

(b) 10-in. wall of block and 4-in. brick

Fig. 4-5. Cavity walls. Also see Figs. 4-11 and 4-12.

backing may be equal to or greater than that of the facing, depending on such structural requirements as wall height and the loads to be carried. Usually the cavity wall is designed so that all the vertical loads are carried by the backing; the outer wall serves as a weather-protective facing. Being tied together, both walls act to resist the wind, although not necessarily equally.

In areas of severe weather exposure, the wall cavity offers three main advantages:

1. It increases the insulating value of the wall and permits use of insulation within the wall.
2. It prohibits the passage of water or moisture across the wall.
3. It prevents the formation of condensation on interior surfaces; therefore, plaster may be applied directly to the masonry without furring or the interior surface may be used as the finished wall without plastering.

Insulation placed within the cavity consists of mats, rigid boards, or non-water-absorbent fill material. Mats or rigid boards may be glass fiber, foamed glass, or foamed plastics. A vapor barrier or dampproofing is required on the cavity face of the inner wall unless waterproofed insulation is used or the insulating rigid boards are held at least 1 in. away from the exterior wall.

Many fill materials absorb water that penetrates into the cavity causing the masonry wall to remain

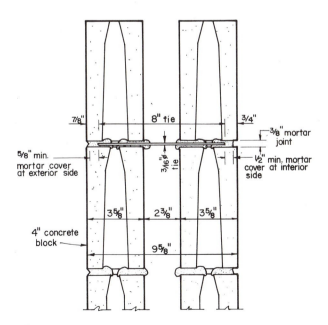

(c) Detailing of metal tie in 10-in. wall

more moist than other walls without fill materials. Increased levels of moisture in a wall can lead to frost damage, efflorescence, and reduced R value.

Walls with insulation should use adjustable ties (Fig. 4-11b). Truss type ties crossing both wythes should be avoided to prevent damaging insulation boards during construction.

Composite Action Walls

A composite action wall (Fig. 4-6) is a multi-wythe wall designed to act as a single member in response to loads. Stress is transferred and shared by the wythes through the mortar or grout filled collar joint and metal ties. Although not the preferred practice, headers can be placed across the filled or unfilled collar joint to provide shear transfer, instead of using metal ties and a filled collar joint.

Anticipated stresses between wythes and collar joints or within headers should not exceed 5 psi for mortared collar joints, 10 psi for grouted collar joints, or the square root of the compressive strength of the header for header construction.

Most types of ties can be used except for Z ties, which should not be used with hollow units. At least one tie for every 2⅔ sq.ft. or 4½ sq.ft. of wall is required for No. 9 gage or ³/₁₆-in.-diameter ties, respectively. Regular ties should be spaced no more than 36 in. horizontally and 24 in. vertically. Adjustable ties should not exceed 1.77 sq.ft. of wall area and should not be spaced more than 16 in. apart.

Header-bonded wythes of solid masonry should have the headers uniformly distributed with a cross-sectional area of at least 4% of the wall surface area. The distance between headers should not exceed 24 in. For hollow masonry, the maximum vertical distance is 34 in. Headers should be embedded at least 3 in. into each wythe. Header construction is not preferred because it does not provide the degree of ductility provided by metal ties. Differential movement also causes the headers to crack. More information on composite-action-wall design is in ACI 530 and ACI 530.1 and the commentaries to these documents.*

Veneered Walls

It is common practice in residential construction to use masonry veneer as a non-load-bearing siding or facing material over a wood frame, as detailed in Appendix A (Fig. A-43). Designed to carry its own weight only, veneer is anchored but not bonded to the backing. In commercial work, the exterior wythe may contain architectural units bonded with joint reinforcement to a concrete masonry backup wall that is loadbearing. If the collar joint is less than 2 in., it is a veneered wall, not a cavity wall.

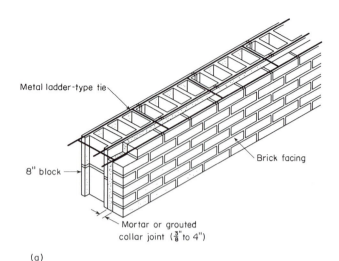

(a)

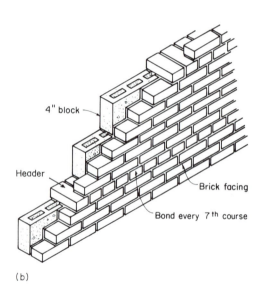

(b)

Fig. 4-6. Multi-wythe composite action masonry walls. (a) Composite action block and brick wall with metal ties. (b) Composite action block and brick wall with headers. Header construction is not preferred as it does not provide as much ductility as metal tied wythes.

The purpose of the veneer is to provide a durable, attractive exterior finish that will reduce or prevent entrance of water or moisture into the building. An air space of at least 1 in. (preferably 2 in.) should be provided between the veneer and the backing to give additional insurance against moisture penetration and heat loss. Flashing and weepholes are provided at the bottom of the air space to eliminate water that may penetrate the veneer.

For residential construction, metal ties anchoring the veneer to the backing are usually 22-gage corrugat-

*See Refs. 75, 76, and 77.

ed, galvanized steel strips ⅞ in. wide. Building code requirements for spacing of such ties vary widely, but an average value would be 16 in. vertically and 32 in. horizontally. For commercial construction unit ties or preferably adjustable ties are used.

Veneer may also be anchored to the backing by grouting it to paperbacked, welded-wire fabric attached directly to the wood studding. The thickness of grout between the backing and the veneer is at least 1 in. No sheathing is required, although it may be added for stiffness, and the need for flashing and weepholes at the base of the wall is eliminated. This type of construction is commonly called reinforced masonry veneer. Other types of veneered walls are illustrated in Figs. 4-7a, b, and c.

Masonry Veneer Over Metal Stud Walls

In recent years a new exterior masonry wall system has appeared on the building scene. It has been widely used but unfortunately it has not always performed successfully (Reference 87).

The system is an adaptation of conventional masonry veneer over wood studs used successfully for one- and two-family dwellings for decades. Generally, the masonry veneers are 4 in. thick and carry no vertical loads from the structure; lateral loads, such as wind, are transferred to the backup system by metal ties anchored in the mortar joints of the veneer. (Fig. 4-7c).

The metal stud, masonry veneer system has some serious design deficiencies. One of the nagging questions is how the stiff masonry veneer can be forced to act with flexible metal studs—structurally, these two elements are incompatible. The stiff veneer will fail by cracking before the flexible metal studs contribute adequate backup support.

Designers are cautioned to carefully analyze the compatibility of the metal studs, which are very flexible, with the masonry veneer, which is very stiff. Laboratory tests indicate that their incompatibility can cause failures.

While the above question still bothers some engineers, many now accept allowable deflection values for the metal stud backup that reflect the behavior of masonry—in the range of L/600 to L/720 for brick (see Reference 88). The National Concrete Masonry Association recommends that concrete masonry veneers over metal studs be designed to prevent cracking of the masonry (see Reference 89).

The Corrosion Question. Structural compatibility is an engineering problem that should be resolved by accepted engineering practice. If it is not, mortar joints in the masonry veneer will fail, allowing water to penetrate the cavity behind the veneer. This raises a second question, that of corrosion.

By looking at the drawing of a typical masonry veneer over metal stud system (Fig. 4-7c), it can be

Fig. 4-7(a). A reinforced concrete masonry wall veneered with brick. The collar joint is not filled with mortar.

Fig. 4-7(b). Block wall veneered with stone.

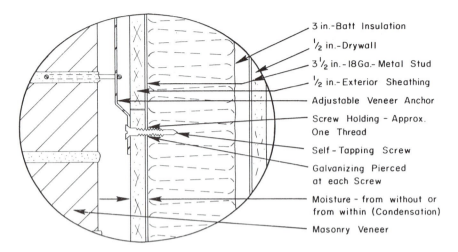

3 in.-Batt Insulation
½ in.-Drywall
3½ in.-18 Ga.-Metal Stud
½ in.-Exterior Sheathing
Adjustable Veneer Anchor
Screw Holding - Approx. One Thread
Self-Tapping Screw
Galvanizing Pierced at each Screw
Moisture - from without or from within (Condensation)
Masonry Veneer

Fig. 4-7(c). Description of a typical masonry veneer over metal stud wall system. The backup support system is light metal framing sheathed with a water-repellent gypsum board facing outside and regular drywall inside. An air space separates the masonry veneer from the support system. Metal anchors or ties attach the veneer to metal studs, which are light-gage C-shaped members screw-attached to U-shaped metal channels running horizontally at top and bottom. Sheathing is attached by driving self-tapping screws through the sheathing into the stud flange. Veneer anchors are screw-attached through the sheathing into the stud flange. The anchors span the air space and are embedded in the mortar joints of the veneer.

seen that all that holds the masonry anchor to the metal stud is a turn or two of the threads of a commonly used self-tapping screw.

There is concern for the likelihood of moisture entering the wall and getting at the screw threads of this anchor system. Unlike the question of relative stiffness of the wall components, this is more a judgement call based on experience.

It can be anticipated that some wind-driven rain will penetrate 4 in. of masonry. In many climates the probability of condensation in the area of the screw threads of the anchor system is high. Whether from without or within, moisture around the pierced galvanized metal stud at the screw location can cause major problems if corrosion occurs.

The late, nationally recognized corrosion expert, Dr. Lewis W. Gleekman, predicted a service life for the system of about seven years when moisture and chlorides are present. That service life is not acceptable.

Lately, some major design firms have been reconsidering the compatibility of masonry veneers over metal stud backup and are returning to the conventional masonry cavity wall or masonry composite wall. This choice is based both on first cost and on life-cycle cost. Design firms are encouraged to make their own critical review of the masonry veneer over metal stud wall system.

Reinforced Masonry Walls

Reinforced concrete masonry walls (Fig. 4-8) are used in cases of high stress concentrations, or in areas of high winds or severe earthquakes. Embedment of steel in grouted vertical and horizontal cores and cavities gives the wall increased strength. This permits the use of higher design stresses and an increase in the distance between lateral supports. Single or multiple wythes may be used.

Single-wythe walls consist of hollow masonry units laid with face-shell mortar bedding. The vertical cores are aligned to form continuous, unobstructed vertical spaces for the reinforcement to be placed and grouted. Two-core block are preferred to three-core block because of the ease in placing reinforcement and grout.

Multi-wythe walls usually consist of two wythes. In two-wythe construction, the wythes are erected with a 1- to 6-in.-wide continuous air space (collar joint) between them, depending on structural requirements,

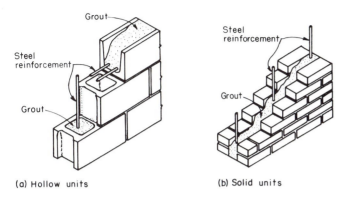

Fig. 4-8. Reinforced walls.

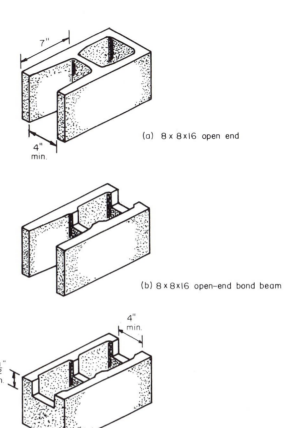

Fig. 4-9. Special units for reinforced walls. Single cell (Fig. 3-6a) and double open-end units (Fig. 1-29c) are also available.

and may have hollow or solid units. Exterior walls may consist of a solid facing and a backing of hollow units, while interior walls may consist of two wythes of hollow units. Reinforcement is placed in the space between the wythes and grouted solid. Depending on site conditions, grouting may be performed intermittently as the wall is erected (low-lift grouting) or completely after erection to a full story-height (high-lift grouting).

Units used in three-wythe construction are usually solid and erected by the low-lift grouting method. The two outer wythes are erected first; reinforcement is placed in the cavity between the wythes; and grout is poured in the cavity. The units of the middle wythe are then "floated" (embedded) into the grout so that ¾ in. of grout surrounds the sides and ends of each unit.

In high-lift grouted construction the grout space is not less than 2 or 3 in. wide, depending on the code. Wythes are tied with No. 9 rectangular ties 4 in. wide and 2 in. narrower than the nominal wall thickness. Spacing of ties is usually 24 in. horizontally and 16 in. vertically for a running bond pattern or 12 in. vertically for a stacked bond pattern. Codes restrict or do not permit the use of kinked or crimped ties in these walls.

Usually vertical and horizontal reinforcement consists of reinforcing steel bars of ⅜-in. minimum diameter. However, building codes permit the use of prefabricated-type continuous joint reinforcement as all or part of the horizontal steel. For Seismic Zones 3 and 4, building codes require that the minimum area of reinforcement in either direction be not less than 0.07% of the gross cross-sectional area of the wall and that the sum of the percentages of horizontal and vertical reinforcement be at least 0.2%.

Special-shape units (Fig. 4-9) have been developed for use in single-wythe reinforced masonry construction. The most common one is the open-end unit shown in Fig. 4-9a. The advantage of this unit is that it can be laid around vertical steel rather than threaded down over the rods. A special H-shaped unit (Fig. 1-29c) also is made for reinforced grouted masonry. It can easily be placed between tall reinforcing bars. In

bond-beam units (Fig. 4-9b and c), low webs not only permit grout to flow both horizontally and vertically, but also facilitate placement of horizontal steel. Some bond-beam units are manufactured with a closed bottom for use in lintel construction; some have depressions in the cross webs to maintain the appropriate steel spacing; and some may also be open-end types (Fig. 4-9b). To permit easy removal of the part of the face shell for cleanout openings at the wall base, some units are scored.

Units used in reinforced concrete masonry sometimes have higher compressive strength than normal load-bearing units; the compressive strength generally ranges from 3,000 to 5,000 psi on the net area. Normally, block producers only stock units that have strengths conforming to ASTM specifications. The local market should be studied for availability of high-strength units.

Grouted Masonry Walls

Grouted masonry walls are similar to reinforced masonry walls but do not contain reinforcement. Grout is sometimes used in load-bearing wall construction to give added strength to hollow walls by filling a portion or all of the cores. It is also used in filling bond beams and, occasionally, the collar joint of a two-wythe wall.

Bonding of Plain Masonry

All of the concrete block or brick in a masonry wall must be bonded (held together) to form a continuous mass. Although several methods may be used, bonding is treated as follows in each of the three directions shown previously (Fig. 4-2):

1. *Horizontally in the plane of the wall* (H). The mortar in the bed joint exerts a shearing bond on the top and bottom of the stretchers,* which overlap to transfer loads and stresses across the head joints. In a stacked bond pattern, some codes require the horizontal joints to be reinforced with the equivalent of a No. 9 gage wire for each 4-in. width of masonry unit at vertical spacings of 16 in. The same reinforcing rules are recommended where there is only a small overlap of one course over the other. The reinforcement is no longer required where 75% of the units in any vertical plane overlap the units below by 1½ in. or half the height of the units, whichever is greater.

2. *Vertically in the plane of the wall* (V). Bonding is achieved by the tensile strength of the mortar and its adhesion to the masonry unit. The force of gravity assists this and helps offset tensile stresses caused by wind or other lateral forces. Because adhesive tensile strength of a mortar joint is only a small fraciton of the compressive strength, careful engineering analysis of all tensile stresses is required, as discussed in Chapter 3.

3. *Horizontally across the width (thickness) of a two-wythe wall* (W). The tensile bond of mortar in a collar joint is not credited; custom and codes require more positive measures such as masonry unit headers or metal ties.

Applicable bond tests include ASTM C952 (Test Method for Bond Strength of Mortar to Masonry Units) and ASTM C1072 (Method for Measurement for Masonry Flexural Bond Strength).

Masonry Headers

Masonry headers consist of stretcher units laid transversely to overlap units of the adjacent wythes or specially shaped header units. Usually they are placed in continuous courses. See the earlier section on "Composite Action Walls" and ACI 530 for header spacing requirements.

Headers have certain disadvantages when compared to metal ties. When headers are used in a bonded

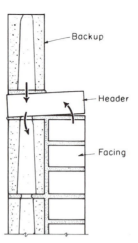

Fig. 4-10. Mortar bond can be impaired by the use of masonry headers.

multi-wythe masonry wall having backup units of concrete masonry and a brick facing, care must be taken to maintain bond between the headers and the mortar. Immediately following the laying of the header course, heavy backup units are set on the brick headers, temporarily loading the headers off center (Fig. 4-10). The mortar below the headers should have time to stiffen or the headers may settle unevenly, causing a fine crack to develop at the external face between the headers and the facing underneath.

Furthermore, backup and facing units may have different thermal and shrinkage properties and consequently different volume changes. Excessive vertical shrinkage of the concrete masonry backup (compared to the facing) may load the headers eccentrically, adding to rupture of the bond between the headers and the mortar at the external face.

The primary advantage of metal ties over masonry headers is their flexibility to accommodate the differential movements between adjacent wythes, thus relieving stresses and preventing cracking. Metal ties are recommended when resistance to rain penetration is important or where wide differences in the physical characteristics of facing and backup exist.

Unit Metal Ties

These ties consist of corrosion-resistant wire embedded in the horizontal mortar joints and engaging all wythes. They are usually made of galvanized steel, but they may be made of stainless steel.

In cavity walls the codes generally require that metal ties be made with ³⁄₁₆-in.-diameter steel wire or 9 gauge joint reinforcement of equivalent strength and stiffness.

---

*A stretcher is a masonry unit laid with its greatest dimension horizontal and its face parallel to the wall face.

Some unit metal ties used with cavity walls used to have a crimp (often called a "drip") located in the center of the tie, as shown in Fig. 4-11a. Its function was to cause any water that finds its way into the cavity to drip off at the crimp before reaching the inner wythe. According to a National Bureau of Standards test report,* the tensile or compressive strength of a ³/₁₆-in.-diameter straight tie in a cavity wall exceeds 1,200 lb. when the mortar has a cube strength of 1,330 psi. A crimp in the tie reduces the strength about 50%. Because crimps reduce the buckling strength of ties, they are not recommended. In high-lift grouted reinforced walls the crimp is not allowed. To maintain the drip feature in straight ties without affecting tie strength, a plastic disk may be installed. Adjustable ties may also induce a drip.

Fig. 4-11a also shows a few commercial tie sizes. Ties used with solid masonry are sometimes bent in a "Z"

shape with 2-in.-long legs at 90 deg. Codes do not allow the Z shape for ties used with hollow masonry; they are usually bent to a rectangular shape 4 in. wide. Tie length is such that the ends are embedded in the face-shell mortar beds at the outside faces of the wall (see Fig. 4-5).

In cases where commercial tie sizes will not fit properly in the outer face-shell mortar beds, a mortar bed base is provided by strips of metal lath or fiberglass/polypropylene mesh laid in the joints receiving the ties; otherwise ties are placed over the head joints or over the webs. A cover of mortar at least ⅝ in. thick is required at any face exposed to the weather (Fig. 4-5c). Ties are spaced not more than 3 ft. horizontally and 24 in. vertically, with each level staggered from the

_____
*See Ref. 19.

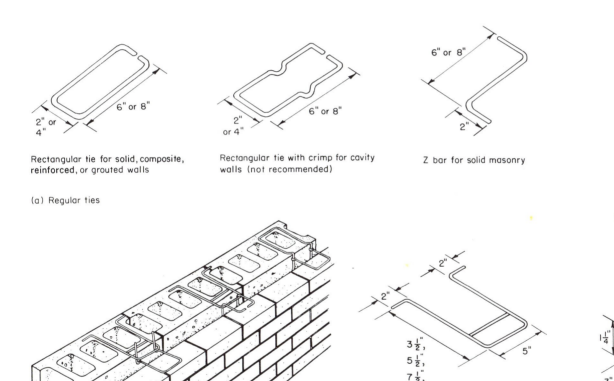

(a) Regular ties

Rectangular tie for solid, composite, reinforced, or grouted walls

Rectangular tie with crimp for cavity walls (not recommended)

Z bar for solid masonry

(b) Adjustable ties

Fig. 4-11. Unit metal ties.

level above or below. The maximum wall surface area per cross wire (9 gauge) of joint reinforcement is 2⅔ sq.ft. and for ³/₁₆-in.-dia. wall ties, 4½ sq.ft. To ensure adequate bonding of the wythes at openings or on both sides of control and isolation joints, unit ties are installed around the perimeter within 12 in. of the opening or joint. Although most rectangular and other ties can be used, Z ties can be used only with solid masonry units, not with hollow masonry units.

Unit metal ties also may be of the adjustable type, as shown in Figs. 4-11b and 4-12f and g. These ties simplify the erection of multi-wythe walls by allowing the mason to erect the wythes independently instead of simultaneously, as with ordinary ties. They also permit adjustment for differences in level between courses. The eye or loop section of the adjustable tie is installed in the first wythe erected, preferably at a head joint or cross web in hollow masonry, with the eyelets or loop close to the face of the wythe. The pintle section, which is always installed in the second wythe, is inserted either up or down.

Adjustable ties for noncomposite action walls must be provided for each 1.77 sq.ft. of wall surface and must be spaced not more than 16 in. horizontally or vertically. The maximum allowable misalignment of bed joints from one wythe to another cannot exceed 1¼ in. for adjustable ties. No more than ¹/₁₆ in. clearance is allowed between connecting parts (eye and pintle) and the pintle must have at least two ³/₁₆-in.-dia. legs.

For noncomposite action walls, the above requirements apply to cavities of 4 in. or less in width.

Continuous Metal Ties

Continuous metal ties are called prefabricated joint reinforcement, mesh or, more commonly, *joint reinforcement*. Consisting of two or more parallel longitudinal wires to which cross wires are welded (Fig. 4-12), joint reinforcement may be used for the following reasons:

1. To act as horizontal reinforcement.
2. To act as longitudinal reinforcement for the control of cracking due to drying shrinkage and temperature changes (discussed later in this chapter).
3. To bond the wythes without using unit metal ties.

See the discussion under "Unit Metal Ties" earlier in this chapter for required spacing of prefabricated joint reinforcement, however vertical spacing of joint reinforcement should not exceed 16 in. (maximum spacing of unit ties is 24 in.).

Joint reinforcement and ties must be corrosion-resistant. Cross wires in joint reinforcement are welded diagonally or perpendicularly to the longitudinal wires, usually at 16-in. spacings. The longitudinals are deformed wire to obtain a better bond with the mortar.

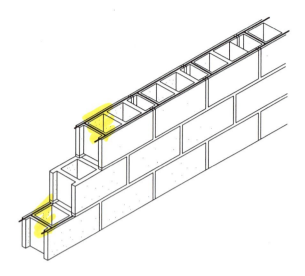

(a) Ladder type joint reinforcement for single-wythe wall

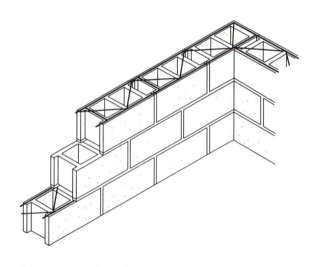

(b) Truss type joint reinforcement for single-wythe wall

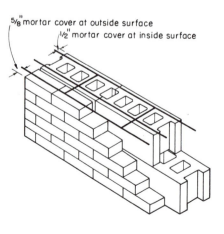

⁵/₈" mortar cover at outside surface
½" mortar cover at inside surface

(c) Ladder tie for multi-wythe wall

Fig. 4-12. Continuous metal ties or joint reinforcement.

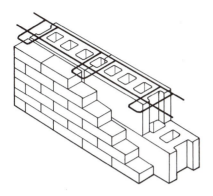

(d) Tab tie for multi-wythe wall

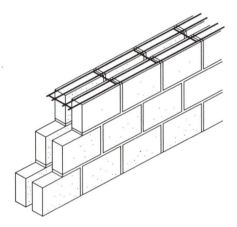

(e) Double ladder tie for multi-wythe wall

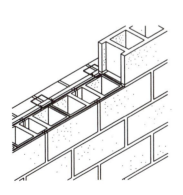

Cross section of adjustable tie for composite action wall

Cross section of adjustable tie for cavity wall (non-composite action)

(f) Adjustable ladder tie

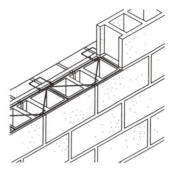
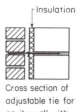

Insulation

Cross section of adjustable tie for cavity wall with rigid insulation

(g) Adjustable truss tie

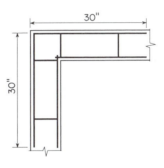

30"

30"

(h) Prefabricated ladder corner

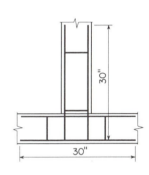

30"

30"

(i) Prefabricated ladder T

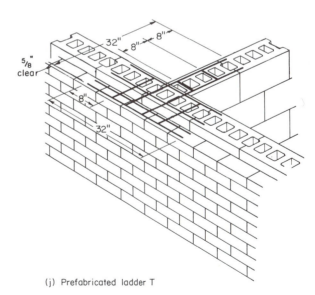

32" 8" 8"

⅝" clear

8"

32"

(j) Prefabricated ladder T

Some manufacturers have developed continuous rectangular-tie assemblies (also called tab ties) that have rectangular ties welded at fixed intervals to longitudinal wires, as shown in Fig. 4-12d.

Joint reinforcement is available in any width, usually 2 in. narrower than the wall, and in commercial lengths of 10 to 12 ft. The longitudinal and cross wires are generally No. 8 or 9 gage; however, $\frac{3}{16}$-in.-diameter wire assemblies are sometimes used. The maximum size of the wire is one-half the mortar joint thickness. Usually No. 9 gage ties are used with $\frac{3}{8}$-in. mortar joints, No. 8 with $\frac{7}{16}$-in. joints, and $\frac{3}{16}$-in. wire with $\frac{1}{2}$-in joints. A mortar cover of at least $\frac{5}{8}$ in. is required at any joint face exposed to the weather (see Fig. 4-5c).

Joint reinforcement in a wall does not guarantee that cracks will not occur. Joint reinforcement does not prevent cracks, it controls the size and spacing of cracks. Cracks occur prior to the joint reinforcement developing any stress.

Wall Patterns

Exposed concrete masonry is an attractive finished wall material for both exteriors and interiors of homes, churches, schools, and public and commercial buildings. One reason for the popularity of concrete masonry is the broad choice of sizes, shapes, textures, and colors. A variety of architectural effects may be obtained by: (1) varying the pattern in which units are laid and (2) applying different treatments to the mortar joints.

For example, if a long, low look is desired, 2-in.-high units 16 in. long will accentuate the horizontal lines (Fig. 4-13). The opposite effect can be achieved by using other size units evenly placed one atop the other to emphasize the vertical lines (stacked bond pattern). Concrete block can also be laid in staggered (running bond), diagonal, and random patterns to produce almost any result the designer may be seeking. Masonry overlapped at least one-fourth the units length is considered running bond by ACI 530.

Customized architectural concrete masonry units have enjoyed an immense popularity not only for use as individual profiles (Fig. 4-14), but to overshadow the mortar pattern (Figs. 4-1 and 4-15). The designer can use his ingenuity to create any of a multitude of pattern arrangements and, by using block with a visible third dimension, an infinite diversity of effects.

In some wall treatments all the joints are accentuated by deep tooling; in others only the horizontal joints are accented. In the latter treatment the vertical joints are tooled, refilled with mortar, and then rubbed flush (after the mortar has partially hardened) to give the joints a texture similar to that of the concrete masonry units. This treatment makes the horizontal joints stand out in relief. It is well suited to walls where strong

Fig. 4-13. Roman concrete brick (*left*) and diagonal stacking of 8x8-in. block (*right*).

Fig. 4-14. The fleur-de-lis in concrete masonry enhance a garden setting.

Fig. 4-15. Customized sculptured block create a rhythmic pattern that overshadows the joint pattern.

horizontal lines are desired. If an especially massive effect is sought, every second or third horizontal or vertical joint can be accented by having all other joints, both horizontal or vertical, refilled with mortar (after tooling) and rubbed flush.

Numerous wall bond patterns are illustrated in Fig. 4-16. Variations of these patterns may be created by projecting or depressing the faces of some units from the overall surface of the wall—or by substituting screen block, split block, or customized architectural units with three-dimensional faces, as discussed in Chapter 1.

Jointing and Control of Cracking

It is well known that various building materials are subject to movement. The movement of concrete masonry walls is due to changes in temperature, changes in moisture content, contraction due to carbonation, and movements of other parts of the structure. When concrete masonry units are bonded together by mortar to form a wall, any restraint that will prevent the wall from expanding or contracting freely will set up stresses within the wall.

Restraint against expansion generally results in low stresses in relation to the strength of the materials and rarely causes damage to concrete masonry walls. Moreover, expansion of the walls is offset by shrinkage from carbonation and drying of the units. Expansion joints are are not necessary in concrete masonry except where required for the length or configuration of the building.* The brick wythe of a multi-wythe wall does require expansion joints.

When contraction of the concrete masonry units is prevented, tensile stresses gradually build up within the wall. If these stresses exceed the tensile strength of the unit, the bond strength between the mortar and the unit, or the shearing strength of the horizontal mortar joint, cracks will occur to relieve the stresses (Fig. 4-17). Cracks usually disfigure the wall and cannot be easily concealed. Also, they affect the lateral stability of adjacent wall sections and must be calked for a weather seal. They can be controlled, as discussed later.

Concrete and clay masonry materials exhibit different properties. Each has its own moisture, thermal, elastic, and plastic flow (creep) properties that must be recognized and taken into account in design, or the masonry wall may not give satisfactory performance.

---

*See Ref. 86.

Fig. 4-16. Forty-two patterns for concrete masonry walls. Some of the patterns illustrated require units of a size or shape that may not be produced in all areas. *Consult local producers for specific available units.*

(a) Running bond, 8×16 in.

(b) Offset bond, 8×16 in.

(c) Offset bond, 8×16 in.

(d) Running bond, 4×16 in.

(e) Offset bond, 4×16 in.

(f) Coursed ashlar, 4×16, 8×16 in.

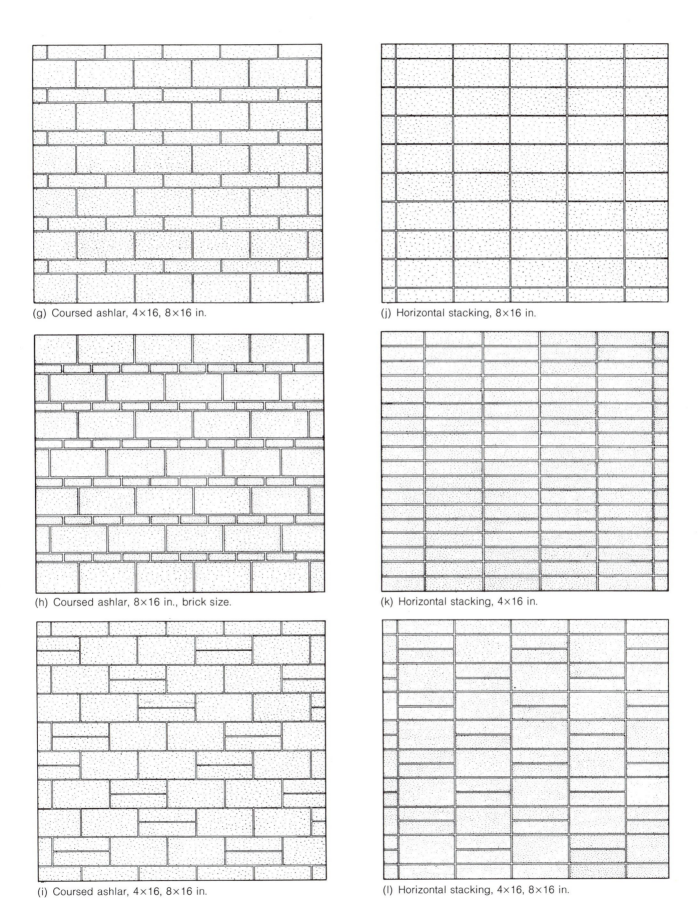

(g) Coursed ashlar, 4×16, 8×16 in.

(j) Horizontal stacking, 8×16 in.

(h) Coursed ashlar, 8×16 in., brick size.

(k) Horizontal stacking, 4×16 in.

(i) Coursed ashlar, 4×16, 8×16 in.

(l) Horizontal stacking, 4×16, 8×16 in.

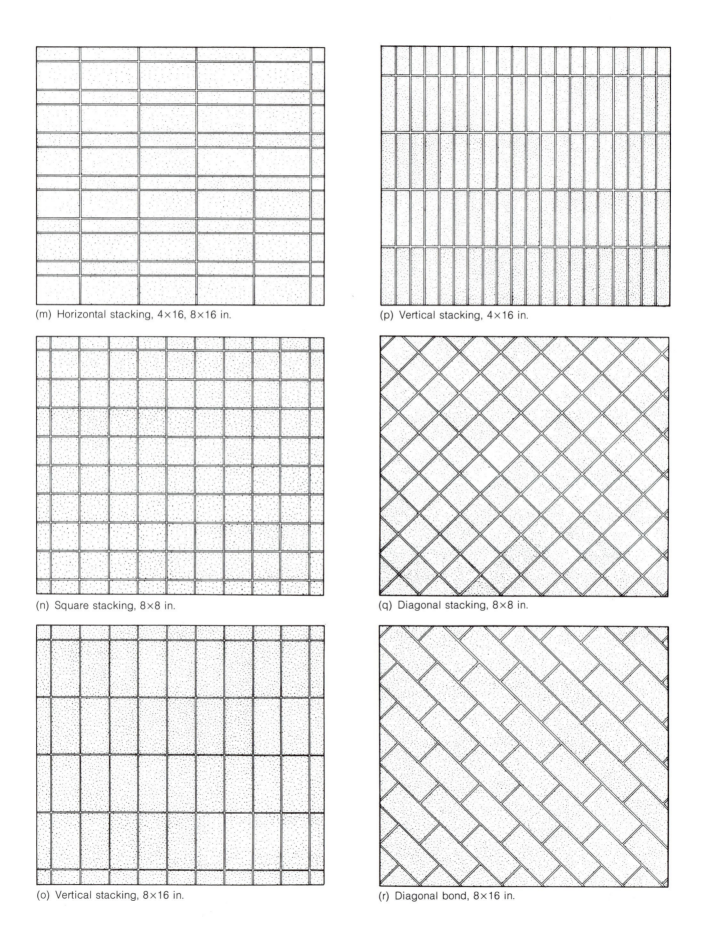

(m) Horizontal stacking, 4×16, 8×16 in.

(p) Vertical stacking, 4×16 in.

(n) Square stacking, 8×8 in.

(q) Diagonal stacking, 8×8 in.

(o) Vertical stacking, 8×16 in.

(r) Diagonal bond, 8×16 in.

(s) Diagonal basket weave, 8×16 in.

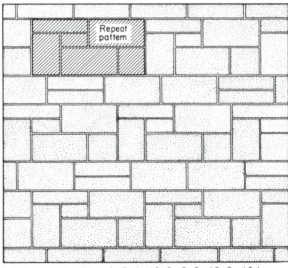

(v) Patterned ashlar, 4×8, 4×12, 4×16, 8×12, 8×16 in.

(t) Diagonal basket weave, 8×16 in.

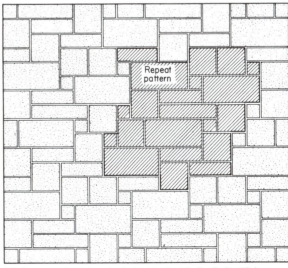

(w) Patterned ashlar, 4×8, 4×16, 8×8, 8×12, 8×16 in.

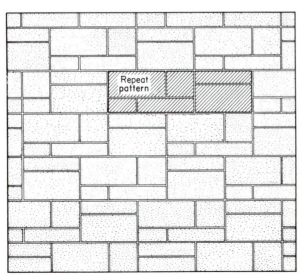

(u) Patterned ashlar, 4×8, 4×16, 8×8, 8×16 in.

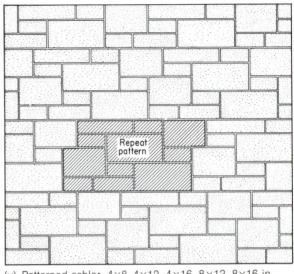

(x) Patterned ashlar, 4×4, 4×12, 4×16, 8×8, 8×16 in.

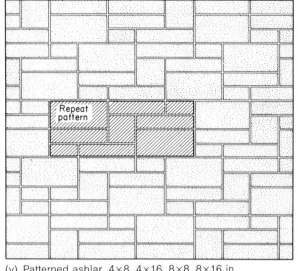

(y) Patterned ashlar, 4×8, 4×16, 8×8, 8×16 in.

(bb) Basket weave, 4×8 in.

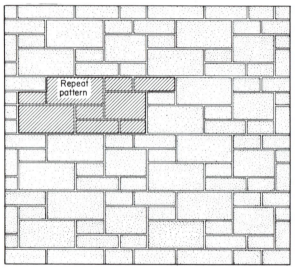

(z) Patterned ashlar, 4×8, 4×12, 8×12, 8×16 in.

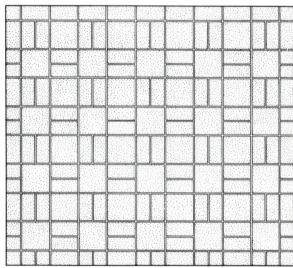

(cc) Basket weave, 4×8, 8×8 in.

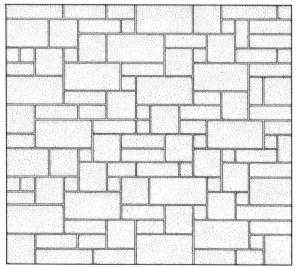

(aa) Random ashlar, 4×4, 4×8, 4×12, 8×8, 8×16 in.

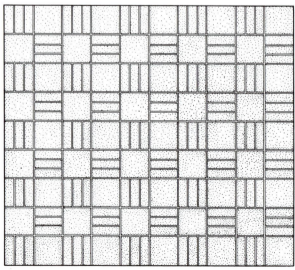

(dd) Basket weave, 8×8 in., brick size.

(ee) Basket weave, brick size.

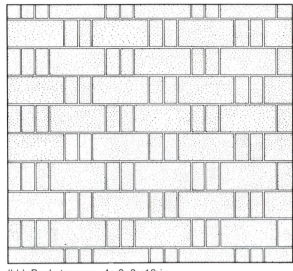

(hh) Basket weave, 4×8, 8×16 in.

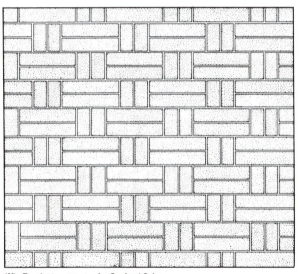

(ff) Basket weave, 4×8, 4×16 in.

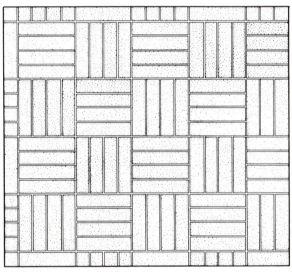

(ii) Basket weave, 4×16 in.

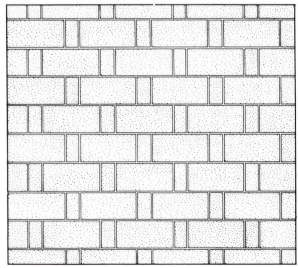

(gg) Basket weave, 4×8, 8×16 in.

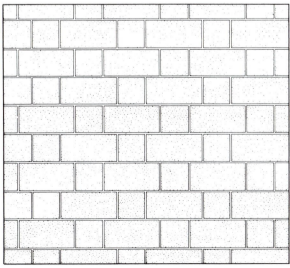

(jj) Basket weave, 8×8, 8×16 in.

(kk) Basket weave, 8×16 in.

(nn) Basket weave, 4×4, 8×16 in.

(ll) Basket weave, 8×16 in.

(oo) Basket weave, 8×8, 8×16 in.

(mm) Basket weave, 4×8, 4×16, 8×16 in.

(pp) Basket weave, 8×8, 8×16 in.

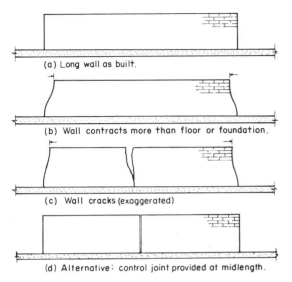

Fig. 4-17. Cracking of long wall by contraction (exaggerated).

Shrinkage Due to Moisture Loss

Building materials, except for the metals, tend to expand with increases in moisture content and contract with drying. For some materials these movements are reversible; for others they are not—or are only partially reversible.

Fire-clay products such as brick expand upon contact with water or humid air, and this expansion is not reversible by drying at normal atmospheric temperatures. Brick has a coefficient of moisture expansion of about 0.03%.

Concrete masonry units such as concrete block expand with a moisture gain and contract with a moisture loss. Of greater immediate concern is the initial drying shrinkage of these units. Many factors affect the volume change of concrete masonry. The drying shrinkage of concrete masonry units is primarily affected by the type of aggregate used, the method of curing, and the method of storage. Standard units made with normal sand and gravel aggregate generally have less shrinkage than those produced with various lightweight aggregates. The difference between the moisture content of the masonry units during construction and after the building is occupied will determine the amount of shrinkage in the wall. Values for shrinkage of concrete masonry can vary from a low of 0.01% to as high as 0.1%. Also see ACI 530.

High-pressure steam curing (autoclaving) of concrete masonry units will reduce shrinkage to approximately one-half the shrinkage of low-pressure (atmospheric) steam-cured units. In pumice units, however, the reduction in shrinkage will be approximately one-third. Few autoclaved units are produced today.

Proper drying of the units before they are laid in the wall will reduce the potential shrinkage of the wall. The degree of dryness will vary according to locality and use. Ideally, the units should be laid with a moisture content corresponding to or preferably slightly below the average annual relative humidity of the outside air for the locality or for the ambient atmospheric condition to which they will be exposed.

As discussed in Chapter 1, ASTM standards classify concrete masonry units as moisture-controlled units or non-moisture-controlled units. Moisture-controlled units have been dried during production and then kept dry so their moisture content does not exceed the requirements spelled out in the standards. Non-moisture-controlled units, as the name implies, have not been subject to limitations on their moisture content during production. However, some codes or standards limit their moisture content at the time of installation to 40% of the total absorption. Moisture-controlled units are not available in some areas and the use of non-moisture-controlled units has been successful with proper design and construction practices. In any case, since moisture content is the predominant factor affecting shrinkage, it is essential that concrete masonry units be kept dry until laid into the wall.

In common construction practice, two methods are used to accommodate shrinkage: minimizing the amount of stress buildup by means of discontinuities in the length of the wall (control joints), and minimizing the width of cracks by means of suitable restraints (joint reinforcement or bond beams). The presence of high-strength-wire joint reinforcement does not eliminate shrinkage, but does distribute the shrinkage stresses and helps control the number and width of drying shrinkage cracks. These two methods can be used separately or together, as will be explained.

Thermal Movements

Theoretically, thermal expansion and contraction movements are reversible if the member is unrestrained. The coefficients for thermal movement can be determined from measurements of length change in prisms by laboratory tests. The generally accepted coefficient for *clay masonry* is 0.04% per 100°F. Thermal movements occur both horizontally and vertically.

Concrete masonry also undergoes external expansion and contraction. The coefficient is 0.045% per 100°F for concrete masonry.

Elastic Deformation

For stresses permitted in *brick masonry,* the relationship between deformation of the structural elements and their stress is approximately linear. The reduction of length of axially loaded masonry elements due to design loads are seldom in themselves critical; however, because these dimensional changes are in addition to those caused by other factors, they must be considered.

Elastic deformations for *concrete masonry* are similar to those found in brick masonry. They are important and must be considered in the design to assure the proper performance of the masonry wall.

Plastic Flow (Creep)

Some materials when continuously stressed gradually yield in the direction of the stress application. This is referred to as creep or plastic flow. In *clay masonry* construction, the clay units themselves are not subject to flow although the mortar joints are. The joints, however, seldom comprise more than 15% to 27% of the volume in compression. A design value of 0.7×10^{-7} per unit of length per pound per square inch is suggested for clay masonry plus mortar.

Creep in *concrete masonry* is much larger than in clay brick construction. For high-strength concrete masonry units, the design value for creep will be somewhat lower than the value for conventional cast-in-place concrete. The ultimate magnitude of creep of plain concrete masonry can be assumed to be 2.5×10^{-7} per unit of length per pound per square inch. Creep is not only a function of stress and time, but is also affected by the physical properties of the concrete and the conditions of exposure.

Composite Walls

The use of clay and concrete units is a popular combination in composite action masonry walls. As pointed out, they respond in different ways to moisture content, temperature, elastic deformation, and plastic flow, with the movement in concrete units greater and generally opposite to that in clay units. Differential movements within the wall can cause cracking when the stresses created by the wall movements exceed the tensile strength of the masonry. The designer can control these stresses and minimize the incidence of cracking by the use of bond breaks, flexible anchorage, and control joints.

Differential Movement

A building is a dynamic structure and its successful performance will depend on the designer's (and builder's) understanding of how all the separate parts interact. The performance of the walls, for example, is dependent on the materials of construction. All of these must be taken into account and a realistic assessment made of their relative effects in service.

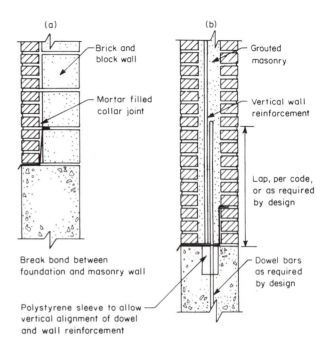

Fig. 4-18. Unbonded and bonded anchorage of masonry wall to foundation.

In the case of cavity walls, any restraints of the differential movement between the exterior and interior wythes could lead to stresses and strains causing distress in the masonry system or lateral deflection (bowing or increased curvature) of the walls. Some type of metal ties connecting the two wythes should be used to accommodate differential movement.

Bond Prevention

Masonry walls are usually supported by concrete foundations. If the walls are of clay brick and the bottom course is fastened to the foundation wall with strong mortar but with no provision for the opposing dimensional movement of the two materials, the result will inevitably be cracking at the foundation corners. In other instances, shrinkage cracks that sometimes originate in the foundation wall can extend up into the masonry wall. These problems can be minimized by a bondbreaker between the foundation and the masonry wall. This can be a layer of building paper or smooth flashing placed under the bottom course of bricks in buildings where it is not necessary to anchor the walls to the foundation, as in Fig. 4-18a. The detail shown in Fig. 4-18b is suggested for grouted masonry walls.

Flexible Anchorage

Masonry walls tied rigidly to the structural frame for lateral support often crack because of differential movements between the two components. These movements can be controlled by flexible anchors that will resist tension and compression but not shear. This flexibility will permit the wall and the structural frame to move independently of each other, within certain limits.

Types of Control Joints

Control joints, also called contraction or movement joints, are continuous, vertical weakened sections built into the wall (Fig. 4-19). If stresses or wall movements are sufficient to crack the wall, the cracks will occur at the control joints and thus be inconspicuous.

A control joint must permit ready movement of the wall in a longitudinal direction and be sealed against vision, sound, and weather. In addition, it may be required to stabilize the wall laterally across the joint.

There are a number of types of control joints for building concrete masonry walls, but the most preferred types are the Michigan, the tongue-and-groove, and the premolded gasket. Fig. 4-20 shows the so-called Michigan type of control joint. It uses conventional flanged units. A strip of building paper is curled into the end core covering the end of the block on one side of the joint and, as the block on the other side of the joint is laid, the core is filled with mortar. The filling bonds to one block but the paper prevents bond to the block on the other side of the control joint. Thus, the control joint permits longitudinal movement of the wall while the mortar plug transmits transverse loads.

Figs. 4-21 and 4-22 show the tongue-and-groove type of control joint. The special units are manufactured in sets consisting of full- and half-length units. The tongue of one special unit fits into the groove of another special unit or into the open end of a regular flanged stretcher. The units are laid in mortar exactly the same as any other masonry units, including mortar in the head joint; this is done so the mason can maintain bond more easily. Also, part of the mortar is allowed to remain in the vertical joint to form a backing against which the calking can be placed. The tongue-and-groove units provide excellent lateral stability for the wall.

Fig. 4-23 shows another type of control joint. It is made by installing a fairly stiff premolded rubber insert in the vertical joint.

Fig. 4-24 illustrates joint stabilizing anchors designed to allow movement at joints while maintaining wall alignment perpendicular to wall movement. These anchors are also adaptable for attaching concrete masonry walls to wood, steel, or concrete framing.

Most of these control joints are first laid up in mortar just the same as any other vertical mortar joint.

Fig. 4-19. Control (contraction) joint in split-block wall. The joint is shown here prior to calking.

Fig. 4-20. Michigan-type control joint.

Fig. 4-21. Special units for tongue-and-groove type of control joint.

Fig. 4-22. Tongue-and-groove units for control joints are made in full- and half-length sizes.

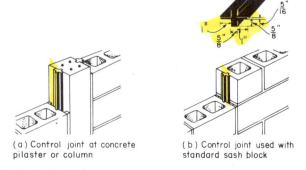

(a) Control joint at concrete pilaster or column

(b) Control joint used with standard sash block

Fig. 4-23. Premolded control joint insert provides lateral support.

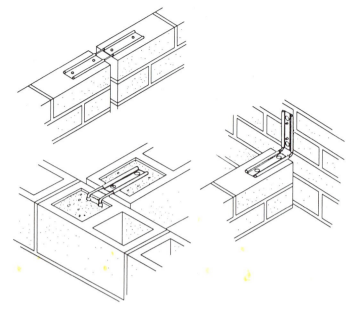

Fig. 4-24. Joint stabilizing anchors allowing shrinkage or expansion through the joint.

Fig. 4-25. At a control joint, mortar is raked out to a depth of 3/8 in. Either a pointing trowel (shown here) or a special wheeled rake may be used.

However, if a control joint is to be exposed to view or the weather, the mortar should be permitted to become quite stiff before a recess is raked out of it to a depth of 3/8 in. (Fig. 4-25). The mortar remaining in the control joint forms a backing to confine an elastomeric joint sealant.

A more preferable technique is to leave the mortar out and use a joint-filler (backer rod) with an elastomeric sealant. Without the mortar, the joint can expand as well as contract. As a remedial technique when joints are inadvertently left out, the joint can be sawed (1/4 in. minimum width) and then filled with backer rod and sealant.

Elastomeric sealants meeting ASTM C920 should be applied in accordance with ASTM C962. Some of the most desirable and successfully used elastomeric sealants for this purpose are the polysulfides, polyurethanes, and silicones. A nonsag or gunnable sealant should be used in joints on vertical surfaces. Tooling is essential to force the sealant into the joint and to match the tooled mortar joints in the masonry (Fig. 4-26). Care must be taken to avoid smearing sealant onto the face of the wall.

Location of Control Joints

No exact rules can be stated for the location of control joints. Each job must be studied individually to determine where joints can be placed without endangering structural integrity. It has been demonstrated in practice that control joints should be not more than 20 ft. apart in exterior walls with frequent openings. In walls without openings the joint spacing may be a little greater but should never be more than 25 ft. to be most effective. A control joint should be located within 10 or 15 ft. of a corner and preferably one header or stretcher unit from the corner. Flexible ties at the corner should be installed to develop the load carrying capacity of the wall, but allow proper movement.

Control joints should also be located at the following points of weakness or high stress concentrations:

1. At all abrupt changes in wall height.
2. At all changes in wall thickness, such as those at pipe or duct chases and those adjacent to columns or pilasters (Figs. 4-27 and 4-28).
3. Above joints in foundations and floors.
4. Below joints in roofs and floors that bear on the wall.
5. At a distance of not over one-half the allowable joint spacing from bonded intersections or corners.
6. At one or both sides of all door and window openings unless other crack control measures are used, such as joint reinforcement or bond beams.

All large openings in walls should be recognized as natural and desirable joint locations. Although some adjustment in the established joint pattern may be required, it is effective to use vertical sides of wall openings as part of the control joint layout. Under windows the joints usually are in line with the sides of the openings. Above doors and windows the joints must be offset to the end of the lintels. To permit movement, the bearing of at least one end of the lintel should be built to slide (Fig. 4-29). Plastic or bituminous sheets or other suitable material should be used for a slip plate.

Openings less than 6 ft. wide require a control joint along one side only, but openings of more than 6 ft. should have joints along both sides (Fig. 4-30). A

Fig. 4-26. Sealant applied to a control joint.

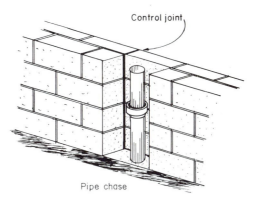

Fig. 4-27. A control joint should be located at a pipe chase or any other abrupt change in wall thickness.

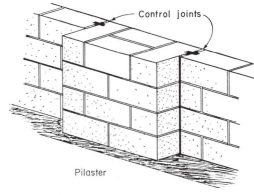

Fig. 4-28. A pilaster edge is a good location for a control joint.

control joint between two windows should be avoided since it will not function properly (Fig. 4-31).

To avoid the occurrence of cracks due to differential movement between concrete masonry and structural framing members, such as columns and pilasters, a space should be allowed between the masonry and

Fig. 4-29. Sliding bearing for a lintel.

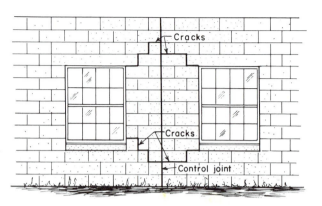

Fig. 4-31. The wrong place for a control joint because cracks may seek a path of less restraint.

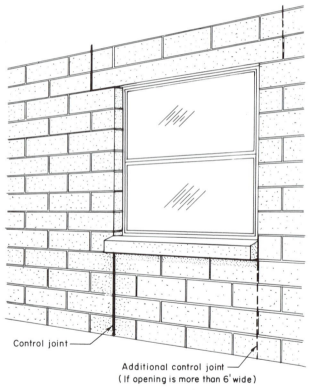

Control joint

Additional control joint
(If opening is more than 6' wide)

Fig. 4-30. Control joints located at window opening to avoid random cracking.

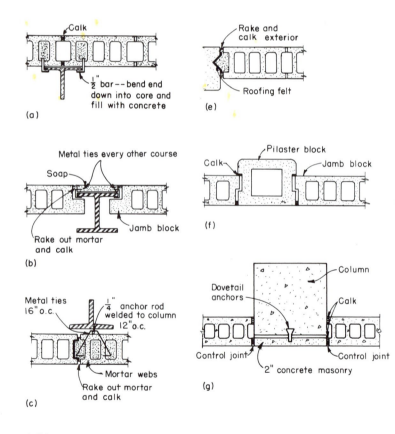

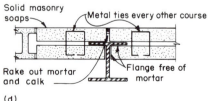

Rake out mortar and calk

Flange free of mortar

(d)

Fig. 4-32. Control joints at columns and pilasters. Joint stabilizing anchors in Fig. 4-24 also are used to attach walls to columns or pilasters while allowing in-plane movement.

member to allow free movement. One or more control joints should be located at the column or pilaster (Fig. 4-32).

When a concrete masonry wall is reduced in thickness across the face of a column, a control joint should be placed along one or both sides of the column. Thin concrete masonry across the column face should be tied to the column by means of dovetail anchors (Fig. 4-32) or another suitable device.

Where bond beams are provided only for crack control, control joints should extend through them. If there is a structural reason for a bond beam, a dummy groove or raked joint should be provided to control the location of the anticipated crack.

A concrete masonry or cast-in-place concrete foundation having both sides backfilled does not usually require control joints. However, long concrete masonry basement walls may require control joints, continuous metal ties (joint reinforcement), or reinforcing bars.

Where concrete masonry units are used as a backup for another material with masonry bond, the control joints should extend through the facing. Control joints need not extend through the facing when using flexible metal ties.

Control joints should extend through plaster applied directly to concrete masonry units. Plaster applied on lath that is furred out from the base requires control joints over previous joints in concrete masonry.

The design, detailing, and spacing of control joints should be by mutual agreement of architect and structural engineer. Both parties should consider: availability of units in project area; engineering aspects as to stress concentrations and requirements for concrete masonry; experience with performance of masonry structures; and esthetics. The engineer should explore alternatives should moisture-controlled units not be available in the project area. Alternatives available include reducing length of wall and adjusting/decreasing joint reinforcment spacing requirements.

Joint Reinforcement

Although concrete masonry walls can be built essentially free of cracks, it is the infrequent crack for which joint reinforcement (Fig. 4-12) is provided. The function of joint reinforcement is not to eliminate cracking in concrete masonry walls but merely to prevent the formation of conspicuous shrinkage cracks. Joint reinforcement does not become effective until the wall begins to crack. After cracking occurs the stresses are transferred to and redistributed by the steel. The result is evenly distributed, very fine cracks that are hardly visible to the naked eye.

The effectiveness of joint reinforcement depends on the type of mortar and the bond between the mortar and the longitudinal wires. The better the bond strength, the more efficient the reinforcement in arresting any cracking. In-service experience has shown that Types M, S, and N mortar should be considered for use with joint reinforcement.

After the joint reinforcement is placed on top of the bare masonry course, the mortar is applied to cover the face shells and joint reinforcement. Minimum mortar cover from the exterior surface to the joint reinforcement should be ⅝ in.; this mortar cover should be ½ in. on the interior as shown in Fig. 4-5c.

Fig. 4-33. Wall with scored block. Joint reinforcement controls drying shrinkage and temperature movements.

Prefabricated or job-fabricated corner and T-type joint reinforcement should be used around corners and to anchor abutting walls and partitions (Fig. 4-12). Prefabricated corners and tees are considered superior because they are more accurately formed, fully welded, and easier to install. A 6-in. lapping of side wires at splices is essential. Continuity of the reinforcement must occur so that tensile stress will be transmitted.

The vertical spacing of joint reinforcement is dependent on the spacing of control joints. In addition, joint reinforcement should be located as follows:

1. In the first and second bed joints immediately above and below wall openings. The reinforcement should extend not less than 24 in. past either side of the opening or to the end of the panel, whichever is less.
2. In the first two or three bed joints above floor level, below roof level, and near the top of the wall.

Joint reinforcement need not be located closer to a bond beam than 24 in. It should not extend through control joints unless specifically called for and detailed in the plans.

Layout of Structural Features

Modular Planning

Modular planning is a method of coordinating the dimensions of various building components to simpli-

Fig. 4-34. Modular planning minimizes the cutting and fitting of block on the job.

Table 4-1. Length of Concrete Masonry Walls by Stretchers

No. of stretchers	Wall length*
1	1'4"
1½	2'0"
2	2'8"
2½	3'4"
3	4'0"
3½	4'8"
4	5'4"
4½	6'0"
5	6'8"
5½	7'4"
6	8'0"
6½	8'8"
7	9'4"
7½	10'0"
8	10'8"
8½	11'4"
9	12'0"
9½	12'8"
10	13'4"
10½	14'0"
11	14'8"
11½	15'4"
12	16'0"
12½	16'8"
13	17'4"
13½	18'0"
14	18'8"
14½	19'4"
15	20'0"
20	26'8"

*Based on concrete blocks 15⅝ in. long and half units 7⅝ in. long, with ⅜-in.-thick head joints. Actual lengths of finished walls are ⅜ in. less than the modular dimensions shown in this table.

fy work and lower the construction cost. Careful planning minimizes cutting and fitting of units on the job, operations that slow up construction (Fig. 4-34). In a modular plan for concrete masonry construction, all horizontal dimensions are given in multiples of half the nominal length of a concrete block, usually 8 in. Vertically the dimensions are given in multiples of the full nominal height of the block.

Tables 4-1 and 4-2 give modular lengths and heights for walls. If necessary, head and bed joints may have different thicknesses.

Door and Window Openings

The example of modular planning given in Fig. 4-35 shows the widths of door and window openings as well as wall lengths in multiples of 8 in. Since the concrete block is produced with dimensions ⅜ in. less than its nominal or modular length of 16 in., the actual dimensions of the finished door and window openings are ⅜ in. greater than their modular dimensions given on the plan. The actual dimension of the finished wall is ⅜ in. less than its modular dimension on the plan. However, the concrete foundation is built to the full modular dimension and theoretically the mason starts the corner masonry unit 3/16 in. in from the end.

Of course, modular design for concrete masonry requires that window and door frames be of the same mode, as shown in Figs. 4-36 and 4-37. The shaded portion of Fig. 4-36 indicates the cutting of units required had nonmodular openings and nonmodular wall length been used.

Corners

An important consideration in modular planning is the method to be adopted for constructing corners. Eight-inch-thick walls do not pose a problem in this regard, but thicker or thinner walls require some attention so that the 4- or 8-in. module is preserved. Figs. 4-39 through 4-43 show some suggested details for handling corner layouts for walls of various thicknesses.

Table 4-2. Height of Concrete Masonry Walls by Courses

No. of courses	Wall height					
	3/8-in. bed joint		7/16-in. bed joint		1/2-in. bed joint	
	8-in. block	4-in. block	8-in. block	4-in. block	8-in. block	4-in. block
1	8″	4″	8 1/16″	4 1/16″	8 1/8″	4 1/8″
2	1′4″	8″	1′4 1/8″	8 1/8″	1′4 1/4″	8 1/4″
3	2′0″	1′0″	2′0 3/16″	1′0 3/16″	2′0 3/8″	1′0 3/8″
4	2′8″	1′4″	2′8 1/4″	1′4 1/4″	2′8 1/2″	1′4 1/2″
5	3′4″	1′8″	3′4 5/16″	1′8 5/16″	3′4 5/8″	1′8 5/8″
6	4′0″	2′0″	4′0 3/8″	2′0 3/8″	4′0 3/4″	2′0 3/4″
7	4′8″	2′4″	4′8 7/16″	2′4 7/16″	4′8 7/8″	2′4 7/8″
8	5′4″	2′8″	5′4 1/2″	2′8 1/2″	5′5″	2′9″
9	6′0″	3′0″	6′0 9/16″	3′0 9/16″	6′1 1/8″	3′1 1/8″
10	6′8″	3′4″	6′8 5/8″	3′4 5/8″	6′9 1/4″	3′5 1/4″
15	10′0″	5′0″	10′0 15/16″	5′0 15/16″	10′1 7/8″	5′1 7/8″
20	13′4″	6′8″	13′5 1/4″	6′9 1/4″	13′6 1/2″	6′10 1/2″
25	16′8″	8′4″	16′9 9/16″	8′5 9/16″	16′11 1/8″	8′7 1/8″
30	20′0″	10′0″	20′1 7/8″	10′1 7/8″	20′3 3/4″	10′3 3/4″
35	23′4″	11′8″	23′6 3/16″	11′10 3/16″	23′8 3/8″	12′0 3/8″
40	26′8″	13′4″	26′10 1/2″	13′6 1/2″	27′1″	13′9″
45	30′0″	15′0″	30′2 13/16″	15′2 13/16″	30′5 5/8″	15′5 5/8″
50	33′4″	16′8″	33′7 1/8″	16′11 1/8″	33′10 1/4″	17′2 1/4″

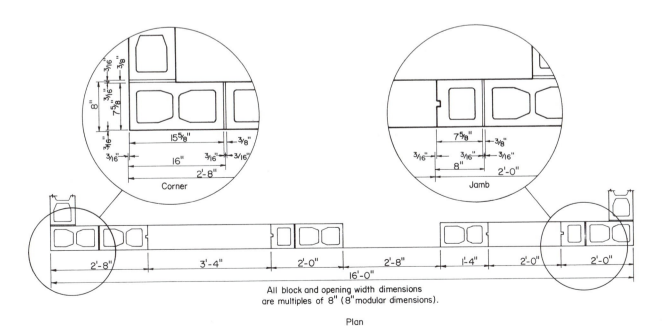

All block and opening width dimensions
are multiples of 8″ (8″ modular dimensions).

Plan

Fig. 4-35. Modular planning of a wall.

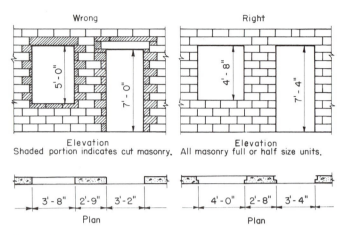

Fig. 4-36. Examples of wrong and right planning of concrete masonry wall openings based on 8x8x16-in. block.

Fig. 4-38. Screen wall enclosing a refuse area. Note the neat corners.

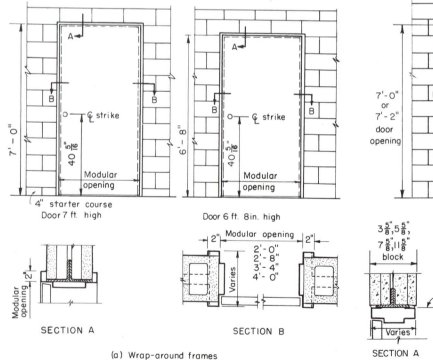

(a) Wrap-around frames

Fig. 4-37. Modular-size door openings.

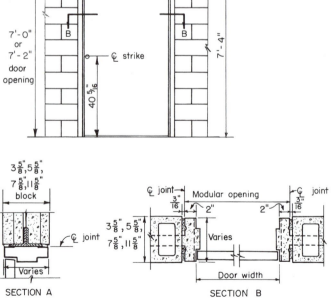

(b) Butt-type frames

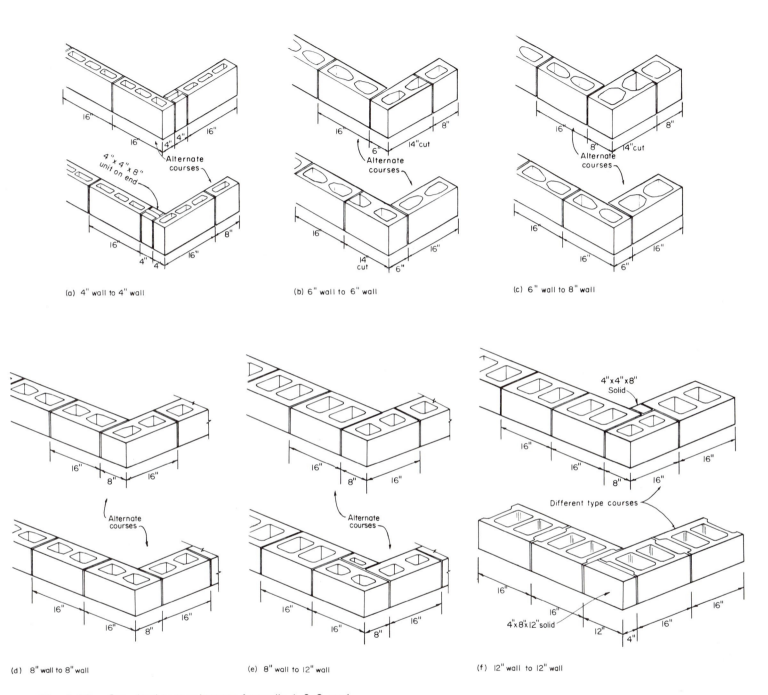

(a) 4" wall to 4" wall

(b) 6" wall to 6" wall

(c) 6" wall to 8" wall

(d) 8" wall to 8" wall

(e) 8" wall to 12" wall

(f) 12" wall to 12" wall

Fig. 4-39. Standard corner layouts for walls 4, 6, 8, and 12 in. thick.

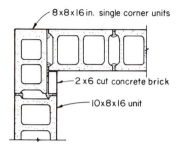

Standard corner construction

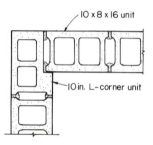

Special L-corner construction

(a) 10" wall to 10" wall

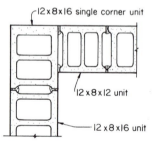

Standard corner construction

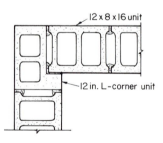

Special L-corner construction

(b) 12" wall to 12" wall

Fig. 4-40. Corner layouts comparing standard and special L units for 10- and 12-in. walls.

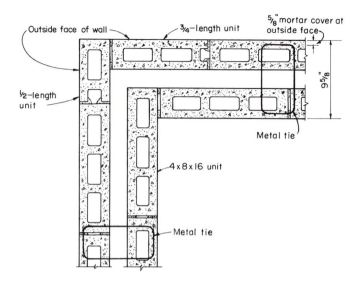

Fig. 4-41. Corner layout for 10-in. cavity wall.

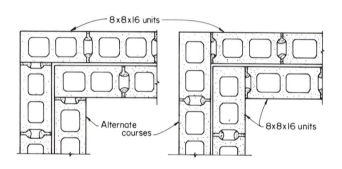

(a) 8" units

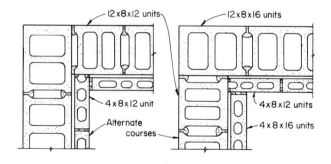

(b) 12" and 4" units

Fig. 4-42. Corner layouts for 16-in. walls.

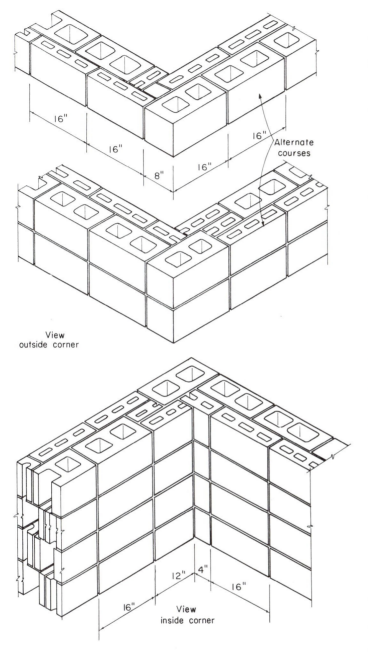

Fig. 4-43. Corner layout for a 12-in. wall with stacked bond pattern.

Intersections

Connection of intersecting walls needs planning unless the method is predetermined by a building code. If there is a choice, two items must be studied: whether the wall will require lateral support and where the control joints will be located. Several conditions are illustrated in Fig. 4-44.

Intersecting load-bearing block walls that depend upon one another for continuity and lateral support, as at A in Fig. 4-44, should be securely anchored to resist all forces that might tend to separate them. Such walls should be connected with true masonry bond so that half of the units of each wall are embedded in the other wall. Alternate connections are shown in Fig. 4-45. Another method is to use the detail of Fig. 4-46a, except that the elastic joint sealant is replaced with a regular mortar joint, as in Fig. 4-47.

At B or C in Fig. 4-44, a full control joint is assumed to be necessary to allow movement in two directions. The joint would be calked with a joint-filler (backer rod) and elastomeric sealant, contain no mortar, and have no block across it.

At the D joint in Fig 4-44, the wall running north-south will try to move transversely to the other. In addition, it requires lateral support—a steel tiebar used as shown in Fig. 4-46a. The appropriate cores are filled with mortar after pieces of metal lath are placed under the cores to support the filling. The ends of the tiebar are embedded in this mortar filling. The result is a hinged control joint that permits the wall to move slightly (at right angles to the abutting wall) and yet have lateral support. The tiebars are spaced not over 4 ft. apart vertically.

In another method for D, a control joint is placed at the junction of the walls. Lateral support is provided by joint reinforcement, strips of metal lath, or ¼-in.-mesh galvanized hardware cloth placed across the joint between the two walls (Fig. 4-45c). When metal strips are used, they are placed in alternate courses in the wall or not more than 18 in. on center. They are sufficiently flexible to permit lateral movement of the abutted wall. If one wall is constructed first, the metal strips are built into it and later tied into the mortar joint of the second wall (Fig. 4-48).

Bond Beams

Bond beams are reinforced courses of block that bond and integrate a concrete masonry wall into a stronger unit. They increase the bending strength of the wall and are particularly needed to resist high winds and hurricane and earthquake forces. In addition, they exert restraint against wall movement and thus reduce the formation of cracks. They may even be used at vertical intervals instead of vertical reinforcing steel.

The bond beams are constructed with special-shape masonry units (Fig. 1-18c through f) filled with concrete or grout and reinforced with embedded steel bars. Bond beams are usually located at the top of walls to stiffen them. Since they have appreciable structural strength, they can be located to serve as lintels over doors and windows. When bond beams are located just above the floor, they act to distribute the wall weight (making the wall a deep beam) and thus help avoid wall cracks if the floor sags. Bond beams may also be located below a window sill. Examples of common

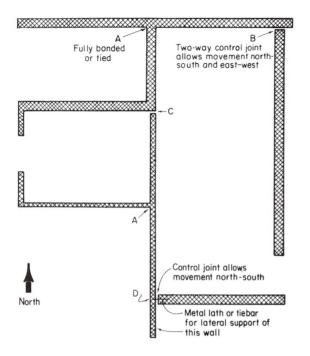

Fig. 4-44. Jointing at intersecting walls.

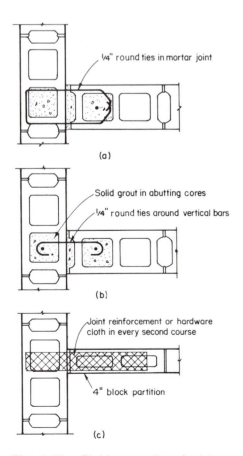

Fig. 4-45. Rigid connections for intersecting walls.

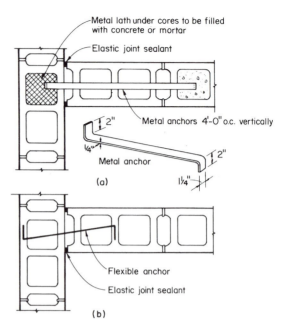

Fig. 4-46. Flexible connections for intersecting walls.

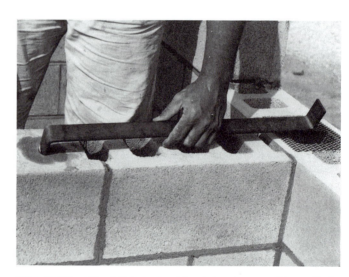

Fig. 4-47. Steel tiebar provides lateral support to wall at right.

bond-beam locations are shown in several figures located in Appendix A.

In load-bearing walls, bond beams are best placed in the top course immediately beneath the roof framing system. Where heads of openings occur within 2 ft. of the roof framing system and/or ceiling in multistory buildings, a bond beam at or immediately above the lintel can be considered the equivalent of a bond beam at the top of the wall or at ceiling height.

In non-load-bearing walls, bond beams may be placed in any one of the top three courses below the roof slab or deck. This permits the bond beam to serve three functions—as a structural tie, lintel, and crack control device.

Reinforcement of bond beams must satisfy structural requirements but should not be less than two No. 4 steel bars in 8-in.-wide bond beams and two No. 5 steel bars in 10- and 12-in.-wide bond beams. In some earthquake-prone areas building codes require a 16-in.-deep bond beam with two additional No. 4 bars located in the top of the beam. Bars should be bent around corners and lapped (Fig. 4-50) according to the local building code.

When the bond beams serve only as a means of crack control, they should be discontinuous at control joints. Where structural considerations require that bond beams be continuous across control joints, a dummy groove should be provided to control the location of the anticipated crack.

If the bond beams are used to replace joint reinforcement, their spacings should be as given in Table 4-3. Note that the area of steel required in the bond beams is greater than that required in the joint reinforcement. This is due not only to a lower reinforcing bar yield strength, but also to a loss of bond-beam effectiveness (assumed to be reduced one-third) because of the wetting effect of grout on the wall and the accompanying increase in ultimate drying shrinkage of the wall.

Fig. 4-48. Hardware cloth provides lateral support to wall at rear.

Table 4-3. Equivalent Spacing of Bond Beams and Joint Reinforcement*

Maximum spacing of joint reinforcement,** in.	Spacing based on bond-beam reinforcement[†]		
	Two No. 4 bars	Two No. 5 bars	Two No. 6 bars
24	8'0"	8'0"	8'0"
16	5'4"	8'0"	8'0"
8	2'8"	4'8"	6'8"

*Adapted from Table 3-3, Ref. 59.
**No. 9 gage wire with yield strength of 65,000 psi
[†]Yield strength of 40,000 psi.

Lintels

Lintels are reinforced horizontal members that span openings in walls; they function as beams to support the weight of the wall and other loads over openings. Concrete masonry lintels may be made up of specially formed lintel masonry units (Fig. 1-18g), bond-beam masonry units, or standard units with depressed, cut-out, or grooved webs. The units are laid end to end to form a channel for placement of reinforcing steel and grout. These lintels are easy to construct, do not re-

Fig. 4-49. Construction of a bond beam for wall reinforcement.

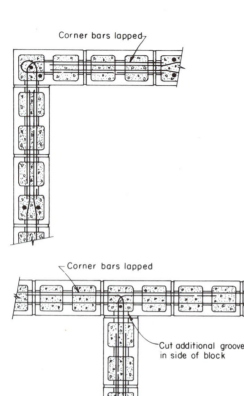

Fig. 4-50. Bond-beam corner details.

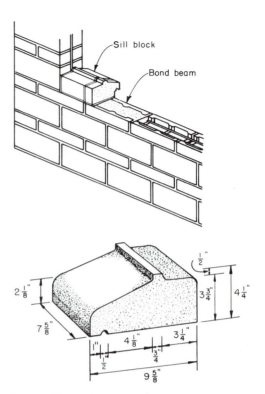

Fig. 4-51. Sill block of modular dimensions.

quire heavy hoisting equipment, and match the bond pattern and surface texture of the surrounding masonry. Lintels may also be conventional precast concrete, steel angles, or concrete masonry members molded on special machines that produce surface textures similar to block.

Lintels should have a minimum bearing of 6 in. at each end. A rough rule of thumb is to provide 1 in. of bearing for every foot of clear span.

Sills

Some concrete masonry producers make sill units in a modular length of 7⅝ in., as shown in Fig. 4-51. They are mortared together to form a continuous sill. Because of the difficulty in ensuring a watertight mortar joint on top of the sill, the course below is constructed of solid units or a bond beam.

Other sill units are shown in Fig. 1-18a.

Piers and Pilasters

Piers are isolated columns of masonry while pilasters are columns or thickened wall sections built contiguous with and forming part of a masonry wall. Pilasters may project on one or both sides of the wall. If the projection is entirely on one side, the pilaster is generally referred to as a flush pilaster or as an interior or exterior pilaster.

Both piers and pilasters are used to support heavy, concentrated vertical roof or floor loads and provide lateral support to the walls. They also offer an economic advantage by permitting construction of higher and thinner walls with plain concrete masonry; otherwise the walls would have to be reinforced or made thicker.

Piers and pilasters may be constructed of special concrete masonry units (Figs. 1-21 and 1-22) or units similar to those used in the wall, whether solid or hollow. Hollow units may be grouted and may or may not contain embedded reinforcement. Grouted piers and pilasters have all vertical joints fully mortared. Pilaster units should be laid with the same mortar quality as is planned for the masonry between pilasters.

Some typical pilaster designs are shown in Figs. 4-53 and 4-54. Most of these layouts are adaptable to piers. Note that grouted piers and pilasters require ties embedded in the face-shell mortar bedding. These ties are necessary to hold the masonry units together while the fresh grout exerts fluid pressure, and to make the

Fig. 4-52. Piers and pilasters accent a wall built of split block laid in an offset bond pattern.

units work together. Tie diameter should not exceed half the joint thickness.

If grouted units are reinforced, it is recommended that they also contain 1/4-in. ties with closed (lapped) ends. These ties should be in contact with the outside of the vertical reinforcing bars. For pilasters such as those in Figs. 4-53 and 4-54, the 1/4-in. ties are installed as the block work progresses.

Planning Weathertight Walls

The outside wythe of a cavity wall can be considered as a rain screen—as are veneered and shingled walls that are backed up with a vented air space. Such walls are highly weather-resistant even in driving rains. On the other hand, solid masonry construction is more vulnerable to leakage. In choosing this type of construction, the designer and builder must accept the possibility and consequences of some leakage or else they must use greater care in selecting materials and overseeing installation. While workmanship is the most important element, it isn't fair to always hold the mason responsible for leaks due to poor workmanship. The owner, architect, and builder share the responsibility because they govern the type of design, materials, and workmanship desired.

The elements of watertight concrete masonry walls are discussed throughout this handbook. They include sound masonry units, proper mortar, and a high standard of construction.

ASTM specifications for concrete brick and block do not specify water permeability, but they do note that protective coatings may be required to prevent water penetration.

Portland cement plaster (stucco) will generally suffice to make a hollow concrete masonry wall weathertight. Acrylic paints and other coatins may also do this, but they may have to be reapplied every few years. Additional information on plaster, paints, and other coatings is given in Chapter 7.

Leaky walls are not confined to any one type of masonry construction. Leaks can occur in walls built of the best materials. The percentage of those that leak is small but receives a disproportionate amount of attention.

Flashing

It is difficult to completely prevent rainwater from entering walls at parapets, copings, sills, projections, recesses, roof intersections, etc., unless proper flashing is installed. In areas subjected to severe driving rains

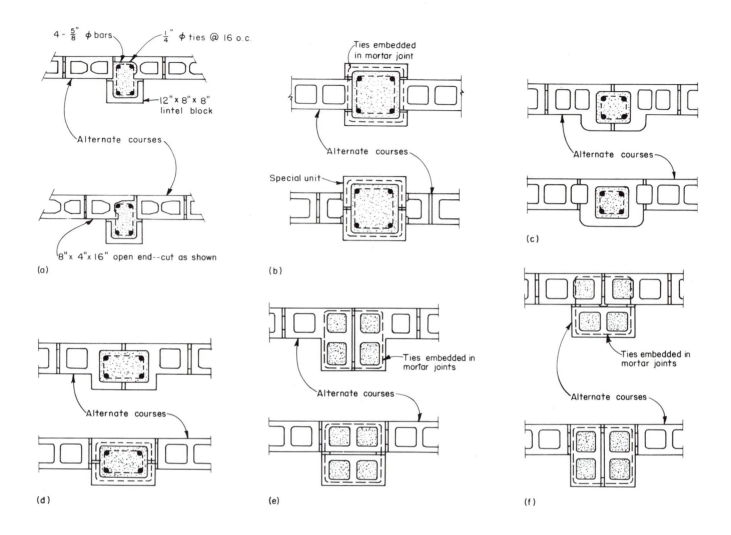

Fig. 4-53. Pilaster layouts.

or where experience has shown that water penetration is to be expected, flashing and weepholes should be provided. Flashing prevents upward movement of water by absorption; thus flashing should interrupt moisture from the ground. Flashing locations are briefly summarized in Table 4-4 and are discussed more thoroughly below. Design details and examples of flashing and weepholes are shown in Figs. 4-55 through 4-60 and in Appendix A.

Any moisture that enters a cavity wall above ground level will gradually travel downward. To divert this water to the exterior of the building, continuous flashing and weepholes are installed at the bottom of a cavity (Fig. A-2, Appendix A).

Where there is a basement and the floor consists of wood joists, the flashing may be located above the bottom of the joists (Fig. A-25, Appendix A). If metal flashing is used, it may be extended at least 2 in. past the inside face of the wall and bend downward at an angle to serve as a termite shield. In past practice, if the flashing was not required to serve as a termite shield, it may have been stopped ½ in. from the outside faces of the wythes; however, current consensus is that all flashing be carried to the face of the wall. In concrete slab-on-ground construction, the flashing extending into the interior wythe may be above the top level of the slab (Fig. A-2, Appendix A).

Flashing should be installed over all windows, doors, and other wall openings not completely protected by overhanging projections. Although flashing may not be required under monolithic sills, it is advisable if sill block are used. Both ends of the sill flashing should be extended beyond the jamb line and turned up at least 1 in. into the wall (Fig. 4-58). Where the underside of the sill does not slope away from the wall or where no drip is provided, the flashing should be extended and

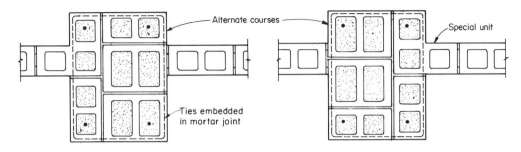

(a) Offset pilaster 24"x 32"

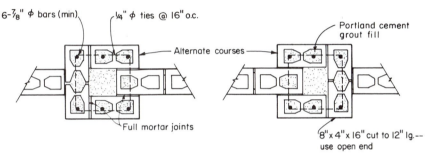

(b) Centered pilaster 24"x 24"

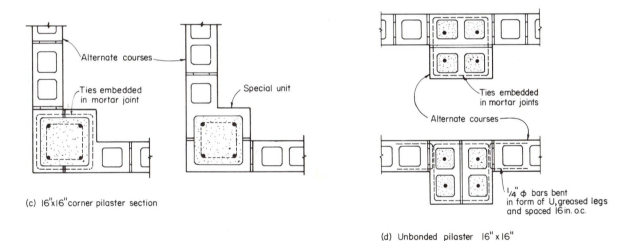

(c) 16"x16" corner pilaster section

(d) Unbonded pilaster 16" x 16"

Fig. 4-54. Other pilaster layouts.

bent down to form a drip. Otherwise, water running down windows and over sills will continue down the face of the building and probably cause unsightly stains.

In structural frame buildings the inner wythe of a cavity wall is constructed flush with and anchored to the beams and columns, while the outer wythe is supported by a steel shelf angle attached to a spandrel beam at each floor level (Fig. A-14b, Appendix A). Flashing normally is necessary even when galvanized or stainless steel angles are used. Flashing should be placed on the shelf angle and extended at least 8 in. up

and over the beam or anchored into a reglet in the beam.

One detail used for flashing at parapets is shown in Fig. 4-59. Others appear in Figs. A-18, A-20, and A-22 in Appendix A.

Suitable flashing materials must be: (1) impervious to moisture penetration; (2) resistant to corrosion caused by exposure either to the atmosphere or to the caustic alkalies that may be present in mortar; (3) sufficiently tough to resist puncture, abrasion, or other damage during installation; and (4) easily formed to

Table 4-4. Flashing Locations and Functions

Location	Function
Under coping	Prevents water penetrating joints in coping stone from entering masonry below
Over window and door heads	Collects and discharges water penetrating from masonry above
Under window sills	Collects and discharges water penetrating sill joints and window jambs
Over foundation	Collects and discharges water penetrating masonry above *and* prevents upward water migration to first course above foundation
Over bond beams	Collects and discharges water from masonry above
Over penetrations	Collects and discharges water from wherever the masonry cavity is interrupted

Flashing of masonry is simplified if the designer and installer realize that *all* masonry imbibes water and retains this moisture until it transpires to the atmosphere through evaporation. Also, under adverse rain exposure, masonry walls may allow water penetration. Water penetration is tolerable only when the penetrating water is intentionally directed away from the building interior by through-the-wall flashing that diverts the water to the exterior. Flashing should be provided in all masonry and should extend beyond the face of the wall.

Flashing should be continuous or made continuous by lapping and bonding the lap joints. Flashing design and installation should always consider end dams at their terminus. An end dam is formed by turning the ends of the flashing upward, thus fabricating a three sided reservoir so water is allowed to exit through weepholes (Fig. 4-58).

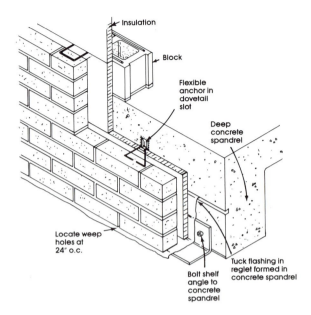

Fig. 4-56. Flashing in brick spandrel supported on steel shelf angle attached to concrete frame. Reference 78.

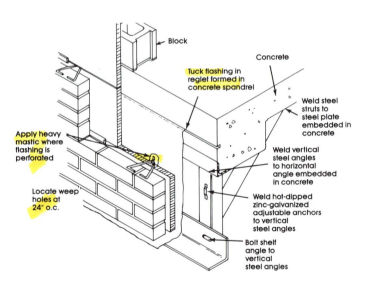

Fig. 4-57. Flashing in brick spandrel supported on steel shelf angle suspended from concrete frame. Reference 78 (also see this reference for flashing details for brick spandrel on steel frame).

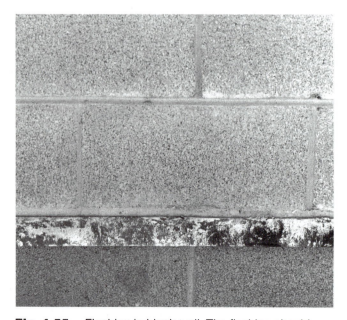

Fig. 4-55. Flashing in block wall. The flashing should extend beyond the face of the wall.

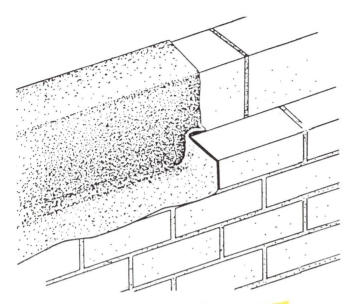

Fig. 4-58. End dam where flashing stops or is not continuous, as at openings in a wall. The flashing should be turned up into the head joint to form a dam that forces water to flow toward the nearest weephole.

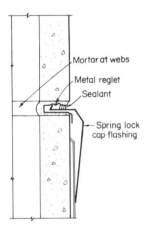

Mortar at webs

Metal reglet

Sealant

Spring lock
cap flashing

Fig. 4-59. Metal reglet used for flashing at parapets.

the desired shape and capable of retaining this shape throughout the life of the structure. The choice of material is governed mainly by cost and suitability. It is advisable to select the type of flashing material carefully since repair and replacement costs will be much higher than the original cost.

Materials generally used for flashing are copper, stainless steel, bituminous fabrics, and plastics. Copper, a durable and easily workable material, has an excellent performance record but is more costly than most other flashing materials. It is also available in special preformed shapes. It does not react with fresh mortar unless chlorides are present. When copper is

Fig. 4-60. Split-face block wall with flashing and rope wick.

exposed to weather, rainwater runoff may stain or discolor the masonry surfaces below. Where this staining or discoloration is objectionable, coated copper or another flashing material should be used.

Stainless steel is durable, highly resistant to corrosion, and workable. Stainless steel flashing is available in several gages and finishes. It will not stain adjacent areas and resists rough handling.

Bituminous fabrics are less costly but also less durable than copper or stainless steel flashing. Care must be exercised during their installation in order to avoid tears and punctures.

Flashing made of plastic materials is also available. However, not all plastics are suitable for use in contact with mortar and thus it is necessary to rely on the past performance of a particular material before selecting it for use in a concrete masonry wall.

Combination flashing consists of materials combined to utilize the best properties of each effectively. Examples of combination flashing are plastic- or asphalt-coated metals, steel- or fiberglass-reinforced bituminous fabrics, copper-plated stainless steel, multi-layered mylar, and fiberglass.

Flashing will reduce the flexural strength of a wall by reducing its continuity in bending resistance and shear. This is not an important factor for houses and small buildings. However, for buildings with tall or thin walls it must be taken into account by the structural designer.

Weepholes

Weepholes are an inseparable companion to flashing; they should be provided immediately above all flashing or other waterstops—except where flashing is located under copings—to drain away any accumulated water (Fig. A-2, Appendix A). The holes are usually located in the head joints of the outer wythe and spaced about 2 ft. apart or preferably installed at multiples of the length of the concrete masonry units. Under adverse weather conditions it may be necessary to install weepholes at the base of the first course at all cell (core) locations in the masonry units. In no case should weepholes be located below grade. They should also be kept small to exclude rodents.

Weepholes are formed by: (1) omitting mortar from part or all of a joint, or (2) placing short lengths of greased or oiled inserts (such as rods, tubing, hose, or cord) into the mortar and extracting them when the mortar is ready for tooling. The inserts should extend up into the cavity for several inches to provide a drainage channel through any mortar droppings that might have accumulated. It is a good practice to fill the cavity with pea gravel to a level just above the weep holes or cell vents. The gravel acts as a "French drain" and also prevents mortar droppings from accidentally filling the cavity at the weepholes.

Whenever possible, the cavity side of weepholes should be covered with copper or plastic insect screen cloth to prevent the entry of insects. Material such as fibrous glass or stainless steel wool may also be placed into the open weepholes. Sometimes absorbent inorganic material is inserted into the holes to act as wicks, drawing moisture out of the cavity. This is especially recommended over lintel or spandrel flashing to prevent the likelihood of staining the wall below. Rope wicks may serve as dams during heavy water penetration but they are excellent for wicking and removing small quantities of water at the base of the cavity or core (Fig. 4-60). Weepholes filled with inorganic materials should be spaced not more than 16 in. on centers.

With proper design and installation, weepholes function as vents to discharge moisture as vapor. When cell vents are used in the bottom of a wall and small holes are used at the top of the cavity, the void is vented, allowing quick removal of moisture from the wall. This technique is getting more popular, especially in colder climates.

Safeguards Against Hurricanes and Earthquakes

To safeguard against high winds, hurricanes, and earthquakes (Seismic Zones No. 3 or 4), the CABO One and Two Family Dwelling Code requires that wall construction be grouted and tied together with reinforcing steel.* Concrete masonry walls should contain reinforcing bars that run horizontally and vertically, extending from the footing to a bond beam at the eave level. Not only should the bars be anchored in the footing and the bond beam, but the cores should be filled in with grout. In addition, roof trusses should be tied down into the bond beam with hurricane clips or steel anchors. See Section R-404.10 of Reference 67 for more details.

———

*See Ref. 67, Sec. R-404.10.

The key to successful and satisfactory construction of concrete masonry in any weather—hot or cold—lies in advanced planning and satisfactory preparation. All-weather construction involves some change in procedures and additional equipment and supplies. The need for these must be anticipated if construction is to be continuous and profitable.

Both hot and cold weather significantly influence the entire masonry construction industry. Hot-weather problems often have been encountered but not recognized, resulting in some sacrifice of quality or increase in construction costs. On the other hand, greater extension of the construction season into the winter months in recent years has resulted in better utilization of manpower and brought to the forefront some techniques of construction not previously well known.

An important part of planning for all-weather construction is accurate weather-forecasting. Builders can plan their construction on the basis of their own weather experience plus information available from the weather bureau. Weather factors important to concrete masonry construction include temperature, wind, rain, snow, humidity, and cloudiness. Combinations of these factors affect construction workers and materials much more seriously than any single factor.

For example, wind and temperature together create a greater impact or chill factor than temperature alone. The cooling effect of a 20-mph wind at 20 above zero (deg. F.) is the same as that of still air at 10 below. Furthermore, a combination of high temperature, low relative humidity, and high wind can cause the early drying of mortar much more rapidly than can one of these elements alone.

Although "normal," "cold," and "hot" are relative terms for masonry construction, "normal" is generally considered as any temperature between 40 and 90 deg. F. Building codes and specifications vary somewhat in this respect. In any case, it should be remembered that some problems may be experienced with such temperatures; for example, those associated with hot weather

may occur even when the temperature is below 90 deg. F.

With modifications of design and construction procedures, concrete masonry construction can be completely satisfactory despite the weather. In many cases concrete masonry construction during hot weather may be little, if any, more expensive than the same construction at normal temperatures; the added cost of masonry construction due to cold weather often amounts to less than 1.5%. As the departure from normal becomes greater, however, the measures necessary to overcome the effects of temperature become more important and more costly.

Hot-Weather Construction

Hot weather poses some special problems for concrete masonry construction. These arise, in general, from higher temperatures of materials and equipment and more rapid evaporation of the water required for cement hydration and curing. Other factors contributing to the problems include wind velocity, relative humidity, and sunshine.

ACI 530.1 defines hot weather construction as occurring when the ambient temperature exceeds 100°F or 90°F with a wind velocity greater than 8 mph (also see Table 5-1).

Masonry Performance at High Temperatures

As the temperature of mortar increases, there are several accompanying changes in its physical properties:

1. Workability is lessened; that is, for a given workability, more water is required.
2. A given amount of air-entraining agent will yield less entrained air.

Table 5-1. Recommendations for All-Weather Masonry Construction*

Air temperature, deg. F.	Construction requirements	
	Materials	Protection
Above 100 or above 90 with wind greater than 8 mph.	Limit open mortar beds to no longer than 4 ft. and set units within one minute of spreading mortar. Store materials in cool or shaded area.	Protect wall from rapid evaporation by covering, fogging, damp curing, or other means.
90 to 40	Normal masonry procedures.	Cover masonry construction with plastic or canvas at end of workday to prevent rain from entering masonry.
Below 40	Heat mixing water. Maintain mortar temperatures between 40 and 120 deg. F. until placed.	Cover masonry construction and materials with plastic or canvas to prevent wetting and freezing for 24 hours.
Below 32	In addition to the above, heat the sand. Frozen sand and frozen wet masonry units must be thawed.	With wind velocities over 15 mph, provide windbreaks during the workday and cover masonry contruction and materials at the end of the workday to prevent wetting and freezing. Maintain masonry above 32 deg. F. by using auxiliary heat or insulated blankets for 24 hours after laying masonry units.
Below 20	In addition to the above, dry masonry units must be heated to 20 deg. F.	Provide enclosures and supply sufficient heat to maintain masonry enclosure above 32 deg. F. for 24 hours after laying masonry units.

*Adapted from guide specifications of the International Masonry Industry All-Weather Council (Ref. 62) and Reference 76. For more specific details, see References 75 to 77.

3. Initial and final set will occur earlier while evaporation will generally be faster.
4. Depending on the surface characteristics, temperature, and moisture content of the concrete masonry units, their absorption of moisture from the mortar will be faster.

As a result of these changes mortar will rapidly lose water needed for hydration. Despite its higher initial water content, mortar will be somewhat more difficult to place and the time available for its use will be shorter.

Early surface drying of mortar joints is particularly harmful. Evaporation removes moisture more rapidly from the outer surface of mortar joints, but the inner parts retain moisture longer and so develop greater strength. A difference of strength across the thickness of the wall reduces the buckling strength of a wall that is concentrically loaded. Also, weak mortar on the surface reduces the strength of the wall under wind and other horizontal loads. For these reasons, during hot weather construction the mortar beds must not be spread more than 4 ft. ahead of the masonry being placed and masonry units must be placed within one minute of spreading the mortar.

Selection and Storage of Materials

During hot weather there is a temptation to reduce the amount of cementitious material in the mortar mixture in order to lessen the heat of hydration released at early ages. Actually, the better solution is to *increase* the amount of cementitious material. This will accelerate rather than retard the mortar's gain in early strength and thus secure maximum possible hydration before water is lost by evaporation.

Mortar materials stored in the sun can become hot enough to significantly affect the temperature of the mortar mixture itself. Covering or shading materials from the sun can be helpful. For example, sand delivered to jobsites normally contains free moisture ranging from 4 to 8%, which is sufficient to ensure that a covered or shaded stockpile of sand remains reasonably cool. If the moisture content drops much below this level, the stockpile should be sprinkled to increase evaporative cooling.

In hot weather the main objective is to see that *all* of the materials of concrete masonry are placed without having acquired excess heat. Heat should be minimized in concrete masonry units by storing them in a

Fig. 5-1. Slump block facing used in an arid region where hot-weather construction procedures are regularly practiced.

cool place and the mortar mixture should be relatively cool. The most effective way of cooling mortar is to use cool water during mixing. Immediately after mortar has been mixed, it begins to rise in temperature and must be protected from further heat gain during construction.

Other Construction Practices

Attention should be given to cooling metal equipment with which the masonry materials, particularly mortar, come into contact. Relatively cool mortar can heat rapidly when transported in a metal wheelbarrow or other container that has been exposed for hours to the sun's rays. Metal mortarboards can become quite hot and wooden ones can become very absorptive in hot weather. Flushing them with water immediately before use and/or working under sunshades can lessen such difficulties.

Since wind and low relative humidities cause increased evaporation, the use of wind screens and fog (water) sprays can effectively reduce the severe effects of hot, dry, windy weather. Also, covering walls immediately after construction will effectively slow the rate of loss of water from masonry. Damp-curing is very effective, particularly in development of tensile bond. If the wall will be subjected to flexure, consideration should be given to damp-curing.

In areas where high ambient air temperatures are common, masonry construction is sometimes rescheduled to avoid hot, midday periods. Construction at night or during the early morning hours can avoid many hot-weather problems.

Cold-Weather Construction

When the ambient temperature falls below normal (40 deg. F.), the productivity and workmanship of masons and the performance of materials may be lowered.

During cold weather masons are concerned not only with their normal construction tasks but also with personal comfort, additional materials preparation and handling, and protection of structures. As temperatures continue to drop, these extra activities consume more time.

ACI 530.1 considers cold weather construction to exist when the ambient temperature falls below 40°F or the temperature of masonry units falls below 40°F.

Masonry Performance at Low Temperatures

Immediately after concrete masonry units are laid during cold weather, several factors come into play. The absorptive masonry units tend to withdraw water from mortar, but mortar, having the property of retentivity, tends to retain water. The surrounding air may chill masonry as well as withdraw water through evaporation. Also, if the masonry units are cold when laid, they will drain heat from mortar. Any combination of these factors influences strength development.

As the ambient temperature falls below normal, mortar ingredients become colder and the heat-liberating reaction between portland cement and water is substantially reduced. Hydration and strength development are minimal at temperatures below freezing. However, construction may proceed at temperatures below freezing if the mortar ingredients are heated. As the ambient temperature decreases, the masonry units should be heated and the structure maintained above freezing during the early hours after construction.

Mortars mixed with cold but unfrozen materials possess plastic properties quite different than those at normal temperatures. The water requirements for a given consistency decrease as the temperature falls, more air is entrained with a given amount of air-entraining agent, and initial and final set are delayed. Also, with lower temperature, the strength gain of mortar is less, although final strength may be as high or higher than that of mortar used and cured at more normal temperatures.

Heated mortar materials produce mortars with performance characteristics identical to those in the normal-temperature range, and thus heating is desirable for cold-weather masonry construction. Mortars mixed to a particular temperature and subjected to a lower ambient level lose heat until they reach the ambient temperature. If the ambient temperature is below freezing when the mortar temperature reaches 32 deg. F., the mortar temperature remains constant until all water in the mortar is frozen. Afterward, the mortar temperature continues to descend until it reaches the level of the ambient temperature.

The rate at which mortar freezes is influenced by the severity of air temperature and wind, the temperature and properties of masonry units, and the temperature of mortar. When fresh mortar freezes, its performance characteristics are affected by many factors, for example, water content, age at freezing, and strength development prior to freezing. Frozen mortar takes on all the outward appearances of hardened mortar, as evidenced by its ability to support loads as well as its ability to bond to surfaces.

Mortar possessing a high water content expands when it freezes and the higher the water content, the greater the expansion. The expansive forces will not be disruptive if moisture in the freezing mortar is below 6%. Therefore, during cold-weather construction every effort should be made to achieve mortar with low water content. Dry masonry units and protective coverings should be used.

Mortar that is allowed to freeze gains very little strength and some permanent damage is certain to occur. If the mortar has been frozen just once at an early age, it may be restored to nearly normal strength by providing favorable curing conditions. However, such mortar is neither as resistant to weathering nor as watertight as mortar that has never been frozen.

Selection of Materials

Cold-weather concrete masonry construction generally requires only a few changes in the mortar mixture. Concrete masonry units used during normal temperatures may be successfully used during cold weather. Under the prevailing recommendations for winter construction (Table 5-1), the masonry units will generally lower the moisture within the mortar to below 6% and so any subsequent accidental freezing should not be disruptive.

At low temperatures mortar performance can be improved—and an early strength gain obtained—by use of Type III high-early-strength cement. Also, mortar made with lime in the dry, hydrated form is preferred to slaked quicklime or lime putty because it requires less water.

Admixtures often considered for use in mortar include antifreezes, accelerators, corrosion-inhibitors, air-entraining agents, and color pigments. Those used with proven success in cold weather are the accelerators and air-entraining agents.

Certain admixtures for mortar are misunderstood in that they accelerate strength gain rather than lower the freezing point. So-called "antifreeze" admixtures, including several types of alcohol, would have to be used in such great quantities to significantly lower the freezing point of mortar that the compressive and bond strengths of masonry would be seriously lowered. Therefore, antifreeze compounds are not recommended for cold-weather masonry construction.

The primary interest in accelerators is to increase rates of early-age-strength development, that is, to hasten hydration of portland cement in mortar. Accelerators include calcium chloride, soluble carbonates, calcium nitrate, calcium nitrite, calcium formate, sodium thiocyanate, silicates and fluosilicates, calcium aluminate, and organic compounds such as triethanolamine.

The most commonly used accelerator in concrete is calcium chloride. However, its use in mortar is prohibited because of possible adverse side effects, such as increased shrinkage, efflorescence, and particularly corrosion of embedded metal. Since calcium chloride may produce corrosion failure, it must not be permitted in mortar for concrete masonry. Where accelerators are desired for use in masonry, nonchloride, noncorrosive accelerators can be used.

The compounds in some proprietary admixtures have been modified to contain corrosion-inhibitors for winter construction of concrete containing embedded metal. Although reports have been published on the performance of corrosion-inhibiting compounds, their value for cold-weather concrete masonry construction has not been fully determined.

Air entrainment increases mortar workability and freeze-thaw durability. The addition of an air-entraining admixture at the mixer on a jobsite is not recommended due to the sensitivity of the admixture and the likelihood of poor control in monitoring air content. Materials with factory controlled amounts of air-entraining agent, such as masonry cement, air-entraining portland cement, or air-entraining lime, should be used if air-entrainment is desired. Masonry cement that entrains air and air-entraining cement perform satisfactorily in mortars used during winter construction.

Some color pigments contain dispersing agents to speed the distribution of color throughout the mortar mixture. The dispersing agents may have a retarding effect on the hydration of portland cement, and this retardation is particularly undesirable in cold-weather masonry construction. In addition, the masonry may have a greater tendency to effloresce.

The use of any admixture in mortar must be approved by the project engineer.

Storage and Heating of Materials

At delivery time all masonry materials should be adequately protected for any exposure conditions at the construction site. During cold weather the safe storage of mortar materials can be accomplished by providing an improvised shelter (Fig. 5-2). As discussed later in this chapter, shelters may be erected by using scaffolding sections, enclosure covers, and lumber. With a properly erected shelter, mortar materials may be delivered, stored, protected from the elements, heated, and mixed within that shelter.

Regardless of the temperature, all masonry materials should be protected from weather effects. Bagged materials and masonry units should be securely wrapped with canvas or polyethylene tarpaulins and stored above the reach of moisture migrating from the ground. The masonry sand should be covered to keep out snow and ice buildup.

The most important consideration in heating materials is that sufficient heat be provided to assure cement hydration in mortar. After all materials are combined, the mortar temperature should be within the range of 40 to 120 deg. F. If the air temperature is falling, a minimum mortar temperature of 70 deg. F. is recommended. Mortar temperatures in excess of 120 deg. F.

Fig. 5-2. A makeshift shelter will protect not only the equipment and materials but also the mixing operation.

Fig. 5-3. Concrete block covered with tarpaulins to keep them dry.

are not recommended; they pose a personnel safety hazard and may cause excessively fast hardening with a resultant loss of compressive and bond strength. Heating requirements for various air temperatures are given in Table 5-1.

Water

When the air temperature drops, water is generally the first material heated for two reasons: it is the easiest material to heat and it can store the most heat, pound

125

Fig. 5-4. A heater pipe extends under the sand from a heater box containing a hot-water tank. Warm sand and a continuous supply of hot water are available with this method.

Fig. 5-5. Sand can be heated over fire in a pipe. Although metal culvert pipe could be used, this heater was made from 30-gal. oil drums (bottoms and tops removed) joined by tack-welding. Sand should be turned over at intervals to avoid scorching.

for pound, of any of the materials in mortar. Recommendations vary as to the highest temperature to which water should be heated. Some specifiers put a maximum of 180 deg. F. on heating water because it poses a personnel safety hazard and there is a possibility of flash set if significantly hotter water comes into contact with cement. Combining sand and water in the mixer first, before adding the cement, will lower the temper-

ature and avoid this difficulty. With this precaution and the use of aggregates that are cold enough, even boiling water may be used successfully, but with extreme caution.

Sand

When the air temperature is below 32 deg. F., sand should be heated so that all frozen lumps are thawed. Generally the temperature of the sand is raised to 45 to 50 deg. F. However, if the need exists and facilities are available, there is no objection to raising the sand temperature much higher; 150 deg. F. is a reasonable upper limit.

Sand should not be heated to a temperature that would cause decomposition or scorching. For example, when sand containing limestone or dolomite is heated to a temperature above 1,200 deg. F., carbon dioxide is liberated; free lime (CaO) and magnesium oxide may then form and cause mortar contamination. When siliceous sand is heated above 1,000 deg. F., scorching can occur. A practical method is to limit the sand temperature by feel (touch of the hand). The stockpile must be mixed periodically to assure uniform heating.

A commonly used method of heating sand is to pile it over a metal pipe containing a fire (Fig. 5-5). Another method is to use steam circulated through coils or injected directly into the sand. Steam boilers are an economical source of heat for winter construction.

An ordinary 40-gal. hot-water heater will raise the temperature of 40 gal. of water about 100 deg. F. in one hour, or the temperature of 1 ton of moist, unfrozen sand about 65 deg. F. in one hour. Frozen sand or water will require more time.

Masonry Units

Masonry units can be heated on pallets in a heated enclosure, preferably at ground level so that they do not constitute a safety hazard to workers below them. If heated enclosures are used for construction, as discussed later, a supply of masonry units can be thawed on the scaffold (Fig. 5-6).

During very cold weather, frozen walls must be heated before grout is poured into cores or cavities. It is recommended that when the air temperature is below 40 deg. F., or it has been below 32 deg. F. during the previous two hours, the air temperature in the bottom of the grout space should be raised above 32 deg. F. before grouting. Heated enclosures may be used for this purpose.

Construction of Temporary Enclosures

The advantages of uninterrupted construction activity during cold weather have persuaded many contractors to build temporary enclosures. Also, U.S. federal regu-

Fig. 5-6. Concrete blocks are placed on the scaffold for enclosure and heating. Corrugated fiberglass panels on independent framework enclose the wall and mason's scaffold.

Fig. 5-8. One floor at a time, this high-rise building is protected from cold and inclement weather to avoid the traditional winter slowdown or shutdown in building construction. Heat and light are provided within the protected portion.

Fig. 5-7. Winter enclosure of masonry construction has many benefits for the contractor, including an earlier completion date and retention of good workers. Masons receive paychecks straight through the cold-weather season instead of only 70% of the year (northern climates).

lations require heated enclosures on government projects to maintain year-around construction.

Providing temporary enclosures for concrete masonry construction has several objectives:

1. To achieve temperatures high enough to facilitate cement hydration in mortar and thus obtain the initial strength required for support of superimposed masonry.
2. To improve the comfort and efficiency of masons and other craftsmen.
3. To protect materials.

Enclosure construction may use temporary, independent framework covered with sheet membranes, tarpaulins, or prefabricated panels (Figs. 5-6 and 5-7). Materials for enclosures include lumber, steel, canvas, building paper, and several plastics such as fiberglass and polyethylene.

Since enclosures are normally built for temporary use, low-cost, lightweight, easily erected and dismantled but tightly sealed enclosure construction is the goal. The type of enclosure constructed depends on whether one or more crafts are to be protected, whether the enclosure is provided by the general contractor or the masonry subcontractor, and whether the building has single or multiple stories. Protection may be given to the walls only, the entire building, or just one floor at a time (Fig. 5-8).

The most popular sheet membrane material is polyethylene film in thickness of 6 to 12 mils (0.006 to 0.012 in.); it is usually clear plastic to let in the light and heat of the sun. Polyethylene reinforced with a fiber or wire mesh is widely used because it is more resistant to ripping. Panels or tarpaulins having reinforced polyethylene with grommets can be used again and again.

A common practice in concrete masonry construction is to enclose only the wall by attaching a protective cover to the mason's scaffold outside. In Fig. 5-9, when the inner wythes of the wall were laid, the enclosure was made by suspending polyethylene tarpaulins from the building framework (permanent or temporary) several feet behind the mason. Of course, prefabricated panels could have been used on the framework. They are made with 1×3- or 2×4-in. wood frames, diagonal bracing or plywood corner gussets, and stapled or

Fig. 5-9. After the weather suddenly turned cold, this scaffolding was enclosed with polyethylene.

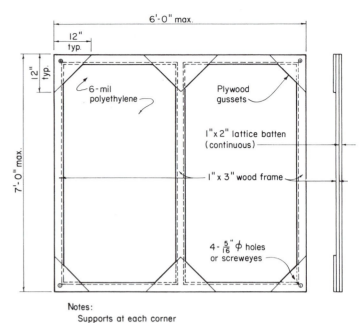

Notes:
  Supports at each corner
  Wind load: 15 psf
  Wood bending stress: 1,200 psi + 33%

Fig. 5-10. Polyethylene enclosure panel.

nailed lattice battens to secure the membrane edges, as shown in Fig. 5-10.

The complete enclosure of a building is usually selected by the general contractor rather than the masonry subcontractor, or is required as described in the job specifications. For low-rise buildings, it offers the opportunity of protection for later craftsmen and for subsequent erection of interior concrete masonry partitions. Generally, low-rise enclosure is accomplished by erecting a temporary framework of timber with polyethylene or canvas cover (Fig. 5-11).

Complete enclosure of a high-rise building begins with the first floor at ground level. Framework, walls, and first floor can be built within this enclosure. The enclosure is then lifted for construction of the second floor and the process repeated until the roof is constructed. As the enclosure is lifted, openings in the newly erected walls are closed with temporary panels until the windows and doors are installed. Hence, the entire building may be heated.

Heating of Enclosures

The most economical and convenient source of heat for temporary enclosures will vary from area to area. Natural gas is often selected for its economy, but other heat sources such as electricity, steam, fuel oil, and bottled propane may be more available. Consideration should be given to the volume of air to be heated and the cost of bringing the source of heat to the construction site.

Fig. 5-11. Tarpaulins form a temporary enclosure when draped over timber framework. Sometimes a mudsill is used to support such framework.

Table 5-2. Heat Loss of Polyethylene Film*

Wind velocity, mph	Heat loss, Btu per sq.ft. per hour		
	Per 1 deg. F.	Per 20 deg. F.	Per 50 deg. F.
0	0.70	14	35
5	0.93	19	46
10	1.05	21	52
20	1.17	23	58
30	1.23	25	61
40	1.26	25	63

*Source: Ref. 29.

Fig. 5-12. A portable oil-fired space heater may be used to heat an enclosure.

Heat loss or gain is calculated in tems of British thermal units (1 Btu = amount of heat required to raise the temperature of 1 lb. of water by 1 deg. F.). Natural gas rates are often expressed in therms (1 therm = 100,000 Btu), and electric rates are based on kilowatt-hours (1 kwh = 3,415 Btu). Portable heaters are classified by Btu of heat output. They are manufactured in various sizes ranging up to several million Btu.

Table 5-2 gives the heat loss of a tight polyethylene enclosure. It does not include heat loss through flaps (masons' entrances), cracks, and loose-fitting film. Thus, the tabulated values should be increased 25 to 50% if they are used as a guide to the size of heater required.

For example, suppose a polyethylene enclosure is 10 ft. wide, 50 ft. long, and 15 ft. high; a 30-mph wind is expected; and enclosure must be warmed 20 deg. F. Heat loss with a tight polyethylene enclosure would be 25 Btu per sq.ft. per hour, and assuming a 50% increase for other heat losses, total heat loss would be 86,000 Btu per hour. The required capacity of a heater must also include the Btu of heat necessary to put heat into the enclosure (3,600 Btu will heat 10,000 cu.ft. 20 deg. F.).

It may be possible to use the heating plant intended for the building if a release is signed for the heating contractor. For a large building it is usually necessary for the builder to supply a temporaray heat source. Where only the wall under construction is enclosed, light portable heaters are applicable (Fig. 5-12).

Although venting ductwork can be connected to some portable heaters, ideally a heater should be located outside the enclosure while blowing hot air in. Venting of fossil-fuel burning heaters is very important because they produce carbon dioxide. Venting such heaters assures a greater supply of fresh air to the workmen, thus maintaining their health and productivity.

It should be noted that excessive *dry* heat will cause rapid drying of the mortar. Thus, live steam is an alternative heat source for enclosures, although it does have limitations; for example, ice may form on the enclosure.

Other Construction Practices

One of the most important practices of concrete masonry work in subfreezing temperatures is to have the masonry units be delivered dry and kept dry until they are laid in the wall. In addition, they should be laid only on a sound, unfrozen surface and never on a snow- or ice-covered base or bed; not only is there danger of movement when the base thaws, but no bond will develop. It is also considered good practice to heat the surface of an existing masonry course to the same temperature as the masonry units to be added. The heat should be sustained long enough to thaw the surface thoroughly.

During cold weather the mortar should be mixed in smaller quantities than usual to avoid excessive cooling before use. Metal mortarboards with built-in electrical heaters may be used if care is taken to avoid overheating or drying the mortar. To avoid premature cooling of heated masonry units, only those units that will be used immediately should be removed from the heat source.

Regardless of temperature, concrete masonry units should never be wetted before being laid in a wall. In cold weather, wetting of masonry units only adds to the problem of keeping masonry units free of moisture that can freeze on the surface. Wetting of units also increases shrinkage and defeats the goal of drawing off water from the mortar to a level below 6% (to prevent mortar expansion upon freezing).

Tooling time during cold weather is less critical than at normal temperatures. Hence, the mason may lay more masonry units before tooling. In all instances, however, the joint should be tooled before mortar gets

Fig. 5-13. Concrete masonry test prisms are covered with polyethylene and cured in a heated box.

Wind forces can make tarpaulins enclosing scaffolds act like sails. Also, the combination of wind forces and dead weight of stacked masonry units can overload enclosure framework unless it has been carefully sized and installed.

too hard. *Caution:* premature tooling at day's end will cause lighter joints.

Upon completion of each section or at the end of each workday, measures should be taken to protect new concrete masonry construction from the weather. During cold-weather construction, rapid drying out or early freezing of the mortar should be prevented. Furthermore, the top of the concrete masonry structure must be protected from rain or snow by plastic or canvas tarpaulins extending at least 2 ft. down all sides of the structure.

During cold-weather construction, careful attention to quality control should be exercised. As shown in Fig. 5-13, masonry test prisms should be covered and stored in a curing box in a manner similar to the treatment given to concrete test cylinders. The temperature inside the box should be maintained between 60 and 80 deg. F. (as required by ASTM C31, Standard Method of Making and Curing Concrete Test Specimens in the Field).

Safety precautions require added emphasis during cold-weather construction for several reasons. Workers not only direct more of their attention to comfort and out-of-the-ordinary construction problems due to low temperature, but they also must overcome personal hazards such as uncertain footing on ice and snow and clumsiness due to bulky protective clothing. Extra care must also be taken with flammable building materials and heaters since the possibility of fire and asphyxiation is increased.

In addition, enclosures for cold-weather protection should be securely guyed or braced if there is danger of snow and ice loads or wind forces. Snow and ice loads can contribute to overloading an *unheated* enclosure.

Chapter 6

CONSTRUCTION TECHNIQUES FOR CONCRETE MASONRY

Masonry structures have been built for thousands of years and the methods passed on from generation to generation by instruction and example. The techniques of concrete masonry construction are well known to architects, engineers, and builders. Recommendations and explanations given in this chapter concern the current state-of-the-art.

Quantity Takeoffs

The first step in construction is to estimate the quantity of materials required. The tables on the next three pages may be used as guides for quantity takeoffs of concrete masonry materials.

For single-wythe walls, wall weights and material quantities are given in Table 6-1. Table 6-2 lists material quantities for composite walls bonded with metal ties or masonry headers. The mortar quantities include the customary allowance for waste that occurs during construction for a variety of reasons. For more information on estimating the quantity of mortar required for a masonry job, see "Estimating Mortar Quantities," Chapter 2.

The breakdown of materials in 1 cu.ft. of mortar is given in Table 6-3. Note that some of the sample mixes given fit the classification of two types of mortar. This is because mortar specifications permit a range in the amounts of lime and sand used for any type of mortar. For all practical purposes, the limits of the range of

Table 6-1. Wall Weights and Material Quantities for Single-Wythe Concrete Masonry Construction*

Nominal wall thickness, in.	Nominal size (width x height x length) of concrete masonry units, in.	Average weight of 100-sq.ft. wall area, lb.**		Material quantities for 100-sq.ft. wall area		Mortar for 100 units, cu.ft.††
		Units made with sand-gravel aggregate†	Units made with lightweight aggregate†	Number of units	Mortar, cu.ft.	
4	4x4x16	4,550	3,550	225	13.5	6.0
6	6x4x16	5,100	3,900	225	13.5	6.0
8	8x4x16	6,000	4,450	225	13.5	6.0
4	4x8x16	4,050	3,000	112.5	8.5	7.5
6	6x8x16	4,600	3,350	112.5	8.5	7.5
8	8x8x16	5,550	3,950	112.5	8.5	7.5
12	12x8x16	7,550	5,200	112.5	8.5	7.5

*Based on ⅜-in. mortar joints.
**Actual weight of 100 sq.ft. of wall can be computed by the formula $WN + 145M$ where W is the actual weight of a single unit; N, the number of units for 100 sq.ft. of wall; and M, the mortar (cu.ft.) for 100 sq.ft. of wall.
†Based on Fig. 1-7, using a concrete density of 138 pcf for units made with sand-gravel aggregate and 87 pcf for lightweight-aggregate units, and the average weight of unit for two- and three-core block.
††With face-shell mortar bedding. Mortar quantities include allowance for waste.

Table 6-2. Material Quantities (Concrete Block, Brick, and Mortar) for 100-Sq.Ft. Area of Composite Walls

Wall thickness, in.	Type of bonding*	No. and size, in., of block		No. of brick	Mortar, cu.ft.**
		Stretchers	Headers		
8	A—metal ties	112.5—4x8x16	—	675	20.0
	B—7th-course headers	97—4x8x16	—	770	12.2
	C—7th-course headers	197—4x5x12	—	770	13.1
12	D—metal ties	112.5—8x8x16	—	675	20.0
	E—7th-course headers	97—8x8x16	—	868	13.5
	F—6th-course headers	57—8x8x16	57—8x8x16	788	13.6

*Key for type shown below.
**Mortar quantities based on ⅜-in. mortar joints with face-shell bedding for the block; mortar quantities include allowance for waste. All unit sizes are nominal.

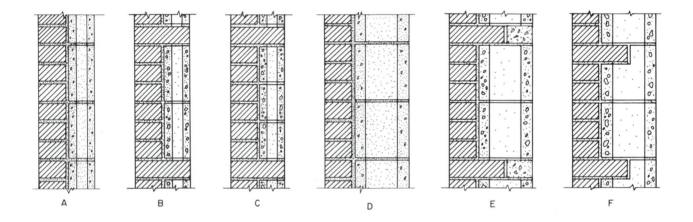

A   B   C   D   E   F

Table 6-3. Sample Quantities of Mortar Materials

Mortar type*	Mix proportions, parts by volume**				Material quantities, cu.ft. for 1 cu.ft. of mortar			
	Portland cement	Masonry cement†	Hydrated lime	Sand	Portland cement	Masonry cement	Hydrated lime	Sand
M	1	1	—	6	0.16	0.16	—	0.97
M or S	1	—	¼	3	0.29	—	0.07	0.96
N	1	—	1	6	0.16	—	0.16	0.97
N or O	—	1	—	3	—	0.33	—	0.99

*See Chapter 2, "Specifications and Types."
**The proportions shown here fall within the range of mixes allowed in Table 2-1, Chapter 2.
†Type N.

Fig. 6-1. Tooling mortar joints of concrete split-block masonry. The fractured block produce walls of rugged appearance.

Table 6-4. Volume of Grout in Two-Wythe Grouted Concrete Brick Walls*

Width of grout space, in.	Grout, cu.yd., for 100-sq.ft. wall area**	Wall area, sq.ft., for 1 cu.yd. of grout**
2.0	0.64	154
2.5	0.79	126
3.0	0.96	105
3.5	1.11	89
4.0	1.27	79
4.5	1.43	70
5.0	1.59	63
5.5	1.75	57
6.0	1.91	53
6.5	2.06	49
7.0	2.22	45

*Adapted from *Volume of Grout Required in Masonry Walls*, Design Aid 15, Masonry Institute of America, Los Angeles, Calif., 1971.
**A 3% allowance has been included for waste and job conditions.

Table 6-5. Volume of Grout in Grouted Concrete Block Walls*

Wall thickness, in.	Spacing of grouted cores, in.	Grout, cu.yd., for 100-sq.ft. wall area**	Wall area, sq.ft., for 1 cu.yd. of grout**
6	All cores grouted	0.79	126
	16	0.40	250
	24	0.28	357
	32	0.22	450
	40	0.19	526
	48	0.17	588
8	All cores grouted	1.26	79
	16	0.74	135
	24	0.58	173
	32	0.49	204
	40	0.44	228
	48	0.39	257
12	All cores grouted	1.99	50
	16	1.18	85
	24	0.91	110
	32	0.76	132
	40	0.70	143
	48	0.64	156

*Adapted from *Volume of Grout Required in Masonry Walls*, Design Aid 15, Masonry Institute of America, Los Angeles, Calif., 1971.
**A 3% allowance has been included for waste and job conditions. All quantities include grout for intermediate and top bond beams in addition to grout for cores.

proportions for the various types of mortar coincide.

Tables 6-4 and 6-5 give grout quantities for grouted walls of concrete brick and concrete block, respectively. These tables are also useful for estimating grout in reinforced concrete masonry.

Construction Procedures

Keeping Units Dry

When delivered to the job, concrete masonry units should be dry enough to comply with specified limitations for moisture content. To be maintained in this dry condition, they should be stockpiled on pallets or other supports free from contact with the ground and then be covered with canvas or polyethylene tarpaulins. The top of a concrete masonry wall should be covered with canvas or polyethylene tarpaulins to prevent rain or snow from entering unit cores during construction.*

Concrete masonry units should never be wetted immediately before or during placement, a practice customary with some other masonry materials. As dis-

---

*See Chapter 5 for hot- and cold-weather construction practices, including coverings and enclosures.

cussed in Chapter 4, when moist concrete masonry units are placed in a wall they will shrink with the loss of moisture. If this shrinkage is restrained, as it usually is, stresses develop that may cause cracks in the walls. Hence, it is important that the units be kept at or dried to at least the moisture content limitations of the applicable specifications.

Sometimes it may be advisable to dry concrete masonry units below the moisture content usually specified for the locality; for example, where walls will be exposed to relatively high temperatures and low humidities in interiors of heated buildings. In such cases it is advisable that, before placement, the units be dried to approximately the average air-dry condition to which the finished construction will be exposed in service.

Damp concrete masonry units can be stacked to facilitate drying and then artifically dried by blowing heated air through the cores and the spaces between the stacked units. An inexpensive and efficient drying method consists of a combination oil- or gas-burning heater and fan. This method of drying works equally well indoors or outdoors and can readily be used in the block plant or at the jobsite.

Mortaring Joints

Two types of mortar bedding are used with concrete masonry: full mortar bedding and face-shell mortar bedding (Fig. 6-2). With full mortar bedding, the unit webs as well as face shells are bedded in mortar. Full bedding is used for laying the first or starting course of block on a footing or foundaiton wall as well as for laying solid units such as concrete brick and solid block. It is also commonly used when building concrete masonry columns, piers, and pilasters that will carry heavy loads. Where some vertical cores are to be solidly grouted, such as in reinforced masonry, the webs around each grouted core are fully mortared. *For all other concrete masonry work with hollow units, it is common practice to use only face-shell bedding.*

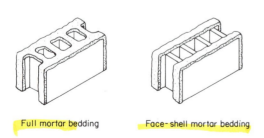

Full mortar bedding   Face-shell mortar bedding

Fig. 6-2. Types of mortar bedding. Only face-shell mortar bedding is ordinarily used to lay concrete block.

Block

For bed (horizontal) joints, all concrete block should be laid with the thicker part of the face shell up. This provides a larger mortar-bedding area and makes the block easier to lift.

A mechnical mortar spreader similar to the device pictured in Fig. 6-3 can be used to speed production, especially on long, straight walls. Unfortunately, the unit shown in Fig. 6-3 is no longer available.

For head joints, mortar is applied only on the face-shell ends of block. Some masons butter (mortar) the vertical ends of the block previously placed; others set the block on one end and butter the other end before laying the block. Time can be saved by placing three or four block on end and then buttering their vertical edges in one operation (Fig. 6-4). Some masons butter both the block already laid and the block to be laid (Fig. 6-5); such application of mortar ensures well-filled head joints.

Regardless of the method used to apply mortar to the vertical edges, each block is brought over its final position and pushed downward into the mortar bed and sideways against the previously laid block so that mortar oozes out of the head and bed joints on both sides of the face shell (Fig. 6-6). This indicates that the joints are well filled.

Caution: Mortar should not be spread too far ahead of the actual laying of units or it will tend to stiffen and lose its plasticity, thereby resulting in poor bond. In hot, dry weather it may be necessary to spread only enough mortar for a few block as they are laid (see Table 5-1 for hot weather requirements).

As each block is laid, excess mortar extruding from the joints is cut off with the trowel and thrown back on the mortarboard for reuse. Some masons apply the extruded mortar to the vertical face shells of the block just laid. If there has been any delay long enough for the extruded mortar to stiffen on the block before it is cut off, it should be reworked on the mortarboard before reuse. Mortar droppings picked up from the scaffold or from the floor should not be reused.

Brick

For concrete brick as well as clay brick, mortar should be spread uniformly thick for the bed joints and furrowed only slightly *if at all* (Fig. 6-7). Many specifications and some building codes prohibit furrowing on bed joints. The weight of the brick and the courses above help compact the mortar and ensure watertight bed joints.

In brick construction, special care should be taken in filling the head joints because they are more vulnerable to water penetration than the bed joints. If head joints are not completely filled with mortar, voids and channels may permit water to penetrate to the inside of the wall. Plenty of mortar should be troweled on the

Fig. 6-3. A mechanical mortar spreader equipped with an electric vibrator applies mortar to the face shells of a block wall (not commercially available).

Fig. 6-4. Most masons spread mortar on the bed joint for only three block at a time and then butter the head joints on them so all three can be placed in rapid sequence. This is convenient because usually the three block plus mortar joints equal 4 ft., the length of a mason's level.

Fig. 6-5. A well-filled head joint results from mortaring both block.

Laying Up a Wall

First Course

end of the brick to be placed so that, when the brick is shoved into place, the mortar will ooze out at the top and around the sides of the head joint, indicating the joint is completely filled (Fig. 6-8). Dabs of mortar spotted on both corners of the brick do not completely fill the head joints, and "slushing" (attempting to fill the joints from above after the brick is placed) cannot be relied on to fill all voids left in the head joints.

On jobs where more than one mason is working, the footing or slab foundation must be level so that each mason can start his section of wall on a common plane and the bed joints will be uniformly straight when the sections are connected. If the foundation is badly out of level, the entire first course should be laid before masons begin work on other courses, or a level plane should be established with a transit or engineer's level.

After the corners are located, masons often string out the masonry units for the first course without mortar in order to check the wall layout (Fig. 6-9). A chalk line is sometimes used to mark the foundation and help align the block accurately.

Before any units are laid, the top surface of the concrete foundation should be clean. Laitance is removed and aggregate exposed by sandblasting, chipping, or scarifying if necessary to ensure a good bond of the masonry to the foundation. Then a full mortar bed is spread and furrowed with a trowel to ensure plenty of mortar along the bottom edges of the block for the first course (Fig. 6-10). If the wall is to be grouted, the mortar bedding for the first course should *not* fill the area under the block cores that are to be grouted; the reason for this is that grout should come into direct contact with the foundation.

The corner block should be laid first and carefully positioned (Fig. 6-11). After three or four block have

Fig. 6-6. Each block is pushed downwards and sideways so that mortar oozes out of the head and bed joints.

Fig. 6-8. Brick is pressed into place so that mortar oozes out of the head and bed joints.

Fig. 6-7. For laying brick, the mason spreads mortar uniformly on the bed joints. Even slight furrowing, as shown here, is sometimes not permitted.

Fig. 6-9. To check the layout, the mason may string out the block without mortar.

Fig. 6-10. A full mortar bed is necessary for laying the first course.

Fig. 6-12. After corners are laid with great care, the mason's level is used as a straightedge to assure correct alignment of the block.

Fig. 6-11. The corner block is laid first and carefully positioned.

Fig. 6-13. Block are leveled by tapping with the trowel handle.

been laid, the mason's level is used as a straightedge to assure correct alignment of the block (Fig. 6-12). Then these units are carefully checked with the level and brought to proper grade (Fig. 6-13) and made plumb (Fig. 6-14) by tapping with the trowel handle. The entire first course of a concrete masonry wall should be laid with such care, making sure each unit is properly aligned, leveled, and plumbed . This will assist masons in laying succeeding courses. Any error at this stage— in the first course—means continuing difficulty in laying up a straight, true wall.

Corners

Corner construction preceeds laying of units between the corners. The corners of a wall normally are built up

Fig. 6-14. The trowel handle is also used to make the block plumb (vertically straight).

higher—four or five courses higher—than the course being laid at the center of the wall. As each course is laid at a corner, it is checked with a level for alignment (Fig. 6-15), for being level (Fig. 6-16), and for being plumb (Fig. 6-17). In addition, each block is carefully checked with a level or straightedge to make certain that the faces of all the block are in the same plane (Fig. 6-18).

Other precautions are necessary at corners to ensure true, straight walls. An accurate method of finding the top of the masonry for each course is provided by the use of a simple story pole made from a 1×2-in. wood strip with markings 8 in. apart (Fig. 6-19); more elaborate metal story poles are commercially available. Also, since each course is staggered one-half unit, the mason easily checks the horizontal spacing of the units by placing his level diagonally across their corners (Fig. 6-20).

Between Corners

After the corners at each end of a wall have been laid up, a mason's line (string line) is stretched tightly from corner to corner for each course and the top outside edge of each block is laid to this line (Fig. 6-21). The line is moved up as each course is laid. Different devices can be used to fasten the line, such as a corner block (Fig. 6-22) held in place at the corner by tension on the line; a line pin (Fig. 6-23) driven into a mortar joint that has set; a line twig (Fig. 6-24) held by a brick and used to eliminate sag in the line; or a line stretcher (Fig. 6-25) fitted over the top of the wall at any convient place.

The manner of handling and gripping a masonry unit is important, and the most practical way for each individual is determined through practice. Generally, the mason tips a block slightly towards himself so that he can see the upper edge of the course below and thus place the lower edge of the block directly over it (Fig. 6-26). By rolling the block slightly to a vertical position and shoving it against the previously laid unit, he can lay the block to the mason's line with minimum adjustment. This speeds the work and reduces the possibility of breaking mortar bond—by not moving the unit excessively after it has been pressed into the mortar. Light tapping with the trowel handle should be the only adjustment necessary to level and align the unit to the mason's line (Fig. 6-27). The use of the mason's level between corners is limited to checking the face of each unit to keep it aligned in a true plane with the face of the wall.

When steel joint reinforcement is required, it is laid on top of the block and mortar is troweled over it (Fig. 6-28).

Caution: All adjustments to final position must be made while the mortar is soft or plastic. Any adjustments made after the mortar has stiffened or even partially stiffened will break the mortar bond and cause cracks between the masonry unit and the mortar.

Fig. 6-15. When laying up corners, the mason checks each course for alignment.

Fig. 6-16. Checking corner for level.

Fig. 6-17. Checking corner for plumb.

Fig. 6-18. Checking the faces of block with a level to make certain all are in the same plane.

Fig. 6-19. Using a story pole to check the spacing of courses.

Fig. 6-20. Diagonal check of the horizontal spacing or stagger of units.

Fig. 6-21. With string line in place, the mason lays block between corners. Masons have individual ways of gripping the block.

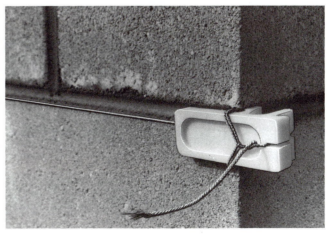

Fig. 6-22. Corner block used to attach string line.

Fig. 6-23. Line pin driven into a hardened mortar joint.

Fig. 6-24. A line twig held in place by a brick eliminates sag in mason's line.

Fig. 6-25. A line stretcher fits over the top of the wall at any convenient spot.

Fig. 6-26. Tipping the block slightly toward himself allows the mason to see the upper edge of the course below so he can place the lower edge of the block directly over the course below. Note the layer of insulation inside the block.

Fig. 6-27. Light tapping brings the block into position with the string line.

Fig. 6-28. Joint reinforcement is placed on the wall and mortar is troweled over it.

This would allow penetration of water. Any unit disturbed after the mortar has stiffened should be removed and relaid with fresh mortar. Realignment of a unit should not be attempted after a higher course has been laid.

Care must be taken to keep the wall surface clean during construction. In removing excess mortar that has oozed out at the joints, the mason must avoid smearing soft mortar onto the face of the unit, especially if the wall is to be left exposed or painted. Numerous embedded mortar smears will detract from the neat appearance of the finished wall. They can never be removed and paint cannot be depended on to hide them.

Any mortar droppings that do stick to the wall should be almost dry before they are removed with a trowel (Fig. 6-29). Then, when dry and hard, most of the remaining mortar can be removed by rubbing with a small piece of concrete masonry (Fig. 6-30) and by brushing (Fig. 6-31).

Closure Unit

Before the closure unit is laid in a course, the length of the opening is checked in order to avoid joints that are too tight or too wide. If necessary, the closure unit is accurately measured, sawed, and dressed for a proper fit in the opening.

When installing the closure unit, the mason butters all edges of the opening and all four vertical edges of the closure unit (Fig. 6-32) before he carefully lowers the unit into place (Fig. 6-33). If any of the mortar falls out, leaving an open joint, the mason removes the closure unit, applies fresh mortar, and repeats the operation. The mortar joints should then be dressed with the point of the mason's trowel (Fig. 6-34). Closure unit locations should be staggered throughout the height of the wall.

Building Composite Walls

In composite action concrete masonry walls the inner wythe laid is parged (backplastered) with mortar not less than ⅜ in. thick before the adjacent wythe is laid (Fig. 6-35). This provides composite action of the two wythes and may help prevent water penetration through the wall. Before parging, however, mortar protruding from the joints of the first-laid wythe should be cut flush while the mortar is still soft (Fig. 6-36); otherwise, parging over the hardened mortar may break the bond in the mortar joints and result in a leaky wall. The use of a plasterer's trowel is far more efficient than a brick trowel for parging.

The preferred practice is to place and parge the inner wythe first before laying the facing wythe. Parging of facing units is usually discouraged. However, if facing is laid first, a mason's level is used during parging to support the facing and prevent it from tilting and

Fig. 6-29. To prevent smears, mortar droppings should be almost dry when cut off the wall with a trowel.

Fig. 6-30. Most of the dry mortar remaining on the wall can be rubbed off with a piece of concrete masonry.

Fig. 6-31. Brushing—the final step in mortar removal.

141

Fig. 6-32. All edges of the opening for the closure block are buttered.

Fig. 6-34. The mortar joints in the closure block are dressed with the point of the trowel.

Fig. 6-33. The closure block is carefully lowered into place.

Fig. 6-35. The first wythe is parged before the adjacent wythe is laid.

breaking bond in the mortar joints (Fig. 6-37). With a light facing wythe, however, it is best to lay up the block backing first so that the parging can be applied on the more substantial wythe (Fig. 6-38).

All solid facing units should be laid with *full* mortar bedding and the head joints completely filled. In header courses the cross joints also should be completely filled; that is, mortar is spread over the entire side of the header unit before it is shoved into place.

For walls greater than 10 in. in thickness, specially shaped header block are sometimes used to bond the facing headers and the backup units with 6th-course bonding, as shown previously in Table 6-2. The special

header block can be laid with the recessed notch up (Fig. 6-39) or down (Fig. 6-40), and with block backup laid in face-shell mortar bedding. In 7th-course bonding (see Table 6-2), the backup consists of stretcher block only, and concrete brick may be used as backup to the brick headers (Fig. 6-41).

Composite walls with header courses are susceptible to water leakage and they have been generally replaced by wire tied walls (Fig. 6-42). When 12-in. block are

Fig. 6-36. If the wall is to be parged, extruded mortar from joints in facing units should be cut off flush before the mortar hardens.

Fig. 6-37. A level is often used during parging to brace the facing.

Fig. 6-38. The block backup is the more substantial wythe on which to apply parging.

Fig. 6-39. Header block can be laid with the notch up.

Fig. 6-40. Here the notch of the header block is down.

Fig. 6-41. Concrete brick can be used to back up the brick headers.

Fig. 6-42. Continuous metal ties (joint reinforcement) are used to tie the two wythes together in this composite wall.

used as backup, two masons might work together to lift each block into place (Fig. 6-43).

Concrete block can also be used as a backup for a stone facing (Fig. 6-44). Because the bed joints in the stone and the block do not line up, corrugated metal strips can be used to tie the two wythes together.

Building Cavity Walls

Cavity walls generally consist of two single-wythe walls that are separated by a continuous air space 2 to 4 in. wide but connected by rigid metal ties embedded in the mortar joints (Figs. 6-45 and 6-46). The ties should have sufficient mortar cover, as shown in Fig. 4-5c.

During construction of the wythes, the cavity must be kept free of mortar droppings that could form a bridge for moisture to pass across the cavity to the interior wythe. A simple method of maintaining a clear cavity is to catch mortar droppings on a board laid across a tier of ties (Fig. 6-47). When the masonry reaches the next level for ties, the board is raised, cleaned, and repositioned (Fig. 6-48).

As an alternate technique, mortar droppings in the cavity may be avoided by spreading the mortar bed so that it remains back about ½ in. from the edge on the cavity. When the next masonry units are laid, the mortar spreads to the edge of the unit without squeezing out into the cavity. Still another method commonly used is to spread the mortar and then draw the trowel over the mortar in an upward and outward direction away from the cavity, thus forming a beveled mortar bed. When the units are laid on such a beveled bed, the mortar spreads only to the cavity edge. In addition, the clean-out opening concept, discussed later in this chapter under "High-Lift Grouting," is also used.

Any mortar fins occuring on the inner faces of the cavity should be trowled flat. This not only prevents the mortar from falling into the cavity but also provides a smooth surface to facilitate placement of insulating materials if required later.

Weepholes are requried at the bottom of cavity walls to shed any rain that penetrates the wall and runs down the inner face of the outer wythe. They should be located in the bottom course at about every second or third head joint in the outside wythe. Cotton sashcord, permanently encased in mortar, makes a good weephole (Fig. 6-49). Flashing (Fig. 6-50) is required in the bottom portion of the cavity wall to direct any water to the weepholes.

During installation of flashing in cavity walls, care must be taken to avoid tearing or puncturing the flashing material, thus destroying its effectiveness. Through-wall flashing can be laid over a thin bed of mortar and then another thin layer of mortar placed on top of the flashing to act as bedding for the next masonry course. If a shelf angle is involved, adhere the flashing to the shelf angle with an adhesive and then a full bed of

Fig. 6-43. Masons work together in handling heavy 12-in. block used as backup in a composite wall.

Fig. 6-45. Cavity wall with unit ties.

Fig. 6-44. Concrete block used as backup for a composite stone-faced wall.

Fig. 6-46. Cavity wall with continuous joint reinforcement, which serves the same purpose as unit ties.

Fig. 6-47. To keep the cavity clean, a wood strip is laid across the ties in the cavity.

Fig. 6-48. The wood strip is lifted to remove mortar droppings.

Fig. 6-49. Cotton sashcord, permanently encased in mortar, makes a good weephole.

Fig. 6-50. Flashing is required in cavity walls to direct water to weepholes.

Plastic flashing, generally joined by heat or appropriate adhesives, does not require expansion seams because it is usually elastic enough to take a certain amount of stretching.*

The cavity face of the interior wythe can also be parged to provide an effective barrier to water penetration. Refer to the earlier section on "Building Composite Walls" for more information on parging.

Building Reinforced Walls

For reinforced masonry wall construction (Fig. 6-51), the procedures used in laying masonry units, placing

mortar can be placed before setting the next masonry course. The seams of the flashing should be thoroughly bonded to ensure continuity of the flashing and prevent penetration of water. Most sheet metal flashing can be soldered, but it requries lockslip joints at intervals to permit thermal expansion and contraction.

---

*Additional information on flashing as well as weephole construction is given at the end of Chapter 4 and in Appendix A.

Fig. 6-51. Reinforced concrete masonry walls are used in earthquake zones.

reinforcing bars, and pouring grout vary with the size of the job, the equipment available, and the preferences of the contractor. Therefore, this section covers only the general requirements of common procedures.

Procedures Before Grouting

Solid or hollow concrete masonry units should be laid so that their alignment forms an unobstructed, continuous series of vertical cores within the wall framework. Spaces in which reinforcement will be placed should be at least 2 in. wide. No grout space should be less than ¾ in. or more than 6 in. wide; if the grout space is wider than 6 in., the wall section should be designed as a reinforced concrete member.

Two-core, plain-end units are preferable to three-core units because the larger cores allow easier placement of reinforcing bars and grout. Also, these units are more easily aligned for their cores to form continuous, vertical spaces. When A-shaped, open-end masonry units (Fig. 4-9) are used, they are arranged so the closed ends are not abutting.

The mortar bed under the first course of block should not fill the core area to be grouted because the grout must come into direct contact with the foundation. All head and bed joints should be filled solidly with mortar for the thickness of the face shell. With plain-end units,

however, it is not necessary to fill the head joint across the full unit width. Also, when the wall is to be grouted intermittently (that is, for reinforcement at 16, 24, 32, or 48 in. on center), only the webs at the extremity of those cores containing grout are mortared. When the wall is to be solidly grouted, none of the cross webs need be mortared since it is desirable for the grout to flow laterally and form the bed joints at all web openings.

Mortar protrusions that cause bridging and thus restrict the flow of grout require an excessive amount of vibration or puddling to assure complete filling of the grout space. Hence, care is necessary that mortar projecting more than ⅜ in. into the grout space be removed and that excess mortar does not extrude and fall into the grout space. The mason can prevent mortar from extruding into the grout space by placing the mortar no closer than ¼ to ½ in. from the edge of the grout space and troweling the mortar bed upward and outward, away from the edge, thus forming a bevel. Mortar droppings in the grout spaces of multi-wythe walls can be caught and removed by using a wood strip as described in the preceding section on cavity wall construction.

Vertical reinforcement may be erected before or after the masonry units are laid. When the reinforcing bars are placed before the units, the use of two-core, open-end, A- or H-shaped units (Figs. 4-9 and 1-29c) becomes desirable in order for the units to be threaded around the reinforcing steel. When the bars are placed after the units, adequate positioning devices are required to prevent displacement during grouting. Both vertical and horizontal reinforcement should be accurately positioned and rigidly secured at intervals by wire ties or spacing devices (Fig. 6-52). The distance between reinforcement and the masonry unit or formed surface must not be less than ¼ in. for fine grout or ½ in. for coarse grout.

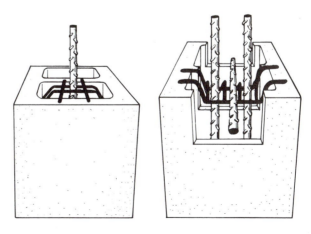

Fig. 6-52. Reinforcing bar spacers.

Horizontal reinforcement is placed as the wall rises. The reinforcing bars are positioned in bond-beam, lintel, or channel units, which are then solidly grouted (Figs. 6-53 and 6-54). Where the wall itself is not to be solidly grouted and cored bond-beam units are used, the grout may be contained over open cores by placing expanded metal lath in the horizontal bed joint before the mortar bed is spread for the bond-beam units. Paper or wood should not be used as a grout barrier because of fire resistance requirements.

To ensure solid grouting of bond beams, it may be necessary to fill those portions of the bond beams between the vertically grouted cores as the bond-beam courses are laid, especially if the spacing of vertically grouted cores is greater than 4 ft. Otherwise, the grout may not flow far enough horizontally from the cores being grouted to completely fill the bond beams.

A concrete masonry wall should be grouted as soon as possible to reduce shrinkage cracking of the joints. However, placing grout before the mortar has been allowed to cure and gain sufficient strength may cause shifting or blowout of the masonry units during the grouting operations. Therefore, to fill the cavity in two-wythe masonry or in large cavities of masonry sections made up or two or more units and containing vertical joints, such as pilaster sections (Figs. 6-55 and 6-56), grout should be poured only after the mortar in the entire height of the masonry has been cured a minimum of 3 days during normal weather or 5 days during cold weather. The hydrostatic or fluid pressure exerted by freshly placed grout on the masonry shell may be ignored when filling hollow-core masonry units, and thus it is necessary to cure mortar in hollow-unit masonry walls for only 24 hours before grouting.

Low-Lift Grouting

Of the two grouting procedures in general use—low- and high-lift* grouting—low-lift grouting is the simplest and most common. This procedures requires no special concrete masonry units or equipment.

Fig. 6-54. Horizontal reinforcing bars positioned in a bond beam that will be solidly grouted.

In low-lift grouting of a single-wythe wall, the wall is built to a height not exceeding 5 ft. before grout is pumped or poured into the cores. This operation is repeated by alternately laying units and grouting at successive heights not exceeding 5 ft. In high-lift grouting, the wall is built to full story-height first before grouting the cores or cavities.

Typical reinforced, single-wythe, hollow masonry construction using low-lift grouting is shown in Fig. 6-53. Vertical cores to be filled should have an unob-

---

*A "lift" is the layer of grout pumped or poured in a single continuous operation. A "pour" is considered to be the entire height of grouting completed in one day; it may be composed of a number of successively placed grout lifts.

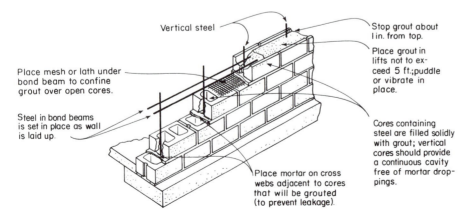

Place mesh or lath under bond beam to confine grout over open cores.

Steel in bond beams is set in place as wall is laid up.

Vertical steel

Place mortar on cross webs adjacent to cores that will be grouted (to prevent leakage).

Stop grout about 1 in. from top.

Place grout in lifts not to exceed 5 ft.; puddle or vibrate in place.

Cores containing steel are filled solidly with grout; vertical cores should provide a continuous cavity free of mortar droppings.

Fig. 6-53. Low-lift grouting of a typical single-wythe reinforced masonry wall.

Fig. 6-55. In this pilaster, a set of four reinforcing bars is connected with ties held in place with soft iron wire.

Fig. 6-56. Large cavities in masonry sections made up of two or more units and containing vertical joints, such as this pilaster, require sufficient time for the mortar to cure before grouting.

structed alignment. Refer to Table 2-8 for minimum grout space dimensions. Also, the vertical reinforcing bars may be relatively short in length because they need to extend only above the top of the lift a moderate distance for sufficient overlap with the reinforcing bars in the next lift. The minimum lap length for bars in tension or compression should be $0.002d_bF_s$ in inches, where d_b is bar diameter and F_s is allowable tensile or compressive stress in reinforcement (ACI 530). Lap length should never be less than 12 in. As an alternate choice, vertical steel may extend to full-wall-height for one-story construction or to ceiling height (plus overlap for splicing) for multistory construction. However, since the longer lengths of steel require the use of open-

end units, some masonry contractors prefer to lap the steel just above each 5-ft. lift.

Grout is handled from the mixer to the point of deposit in the grout space as rapidly as practical (Figs. 6-57, 6-58, and 6-59). Pumping or other placing methods that prevent segregation of the mix and limit grout splatter are used. On small projects, the grout is poured with buckets having spouts or funnels to confine the grout and prevent splashing onto the face or top surface of the masonry. Grouting should be done from the inside face of the wall if the outside will be exposed; dried grout can deface the exposed surface of a wall and be detrimental to the mortar bond of the next masonry course. On most projects, grout pumps are recommended to save time and money.

Fig. 6-57. Ready mixed portland cement grout is delivered by truck mixer into the pump hopper. Note the hose extending to work above.

Fig. 6-58. While helping to handle the grout hose, a laborer controls the pump shutoff with a hand button.

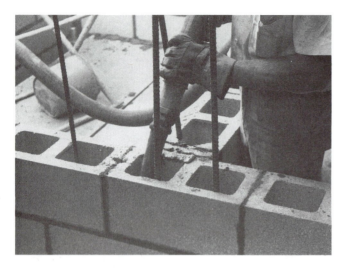

Fig. 6-59. View of grout shooting from the hose into a core.

Fig. 6-60. Grouting is stopped about 1 in. below the top of the block to form a key with the next lift.

Whenever work is stopped for one hour or longer, a horizontal construction joint should be made by stopping the grout pour about 1 in. below the top of the masonry unit to form a key with the next grout lift (Fig. 6-60).

During placement, grout should be rodded (usually with a 1×2-in. wooden stick) or mechanically vibrated to ensure complete filling of the grout space and solid embedment of the reinforcement. It takes very little effort to do this consolidation job properly because of the fluid consistency of the grout. When high-absorption masonry units are used, it may be necessary to rerod or revibrate the grout 15 to 20 minutes after placement; this will overcome any separations of the grout from the reinforcing steel and eliminate voids caused by settlement of the grout and absorption of water from the grout into the surrounding masonry.

Over-vibration, however, must be carefully avoided at this stage; more hazardous than during initial consolidation, it can cause blowouts, broken ties, cracked masonry units, or segregation of the grout.

In low-lift grouting of masonry with two or more wythes, the exterior wythe is laid up a maximum of 18 in. above the interior wythe. After the interior wythe is laid, the cavity betwen the wythes is grouted in lifts not to exceed six times the width of the grout space, with a maximum of 8 in. A minimum mortar curing period is not necessary before grouting. Grout is poured into the grout space to within 1 in. of the top of the interior wythe and then consolidated.

Where there are more than two wythes, the middle wythe (usually of brick size) may be built by "floating" the units in the grout space; that is, pushing the units down into the grout so that a ¾-in. depth of grout surrounds the sides and ends of each unit. No units or pieces of a unit less than 10 cu.in. in size should be embedded in the grout by floating.

High-Lift Grouting

With this procedure, grouting is delayed until the wall has been laid up to full-story-height before grouting commences in lifts not exceeding 5 ft. High-lift grouting is intended for use on wall construction where reinforcement, openings, or masonry unit arrangements do not prevent the free flow of grout or inhibit the use of mechanical vibration to properly consolidate the grout in all cores or horizontal grout spaces. The vertical cores should have an unobstructed alignment with a minimum dimension of 3 in. and a minimum area of 9 sq.in. for a 12-ft. pour of coarse grout (see Table 2-8 for other requirements). In two-wythe masonry the minimum dimension of the grout space (cavity) between wythes is 2 or 3 in., depending on the governing code and pour height (Table 2-8); the maximum is 6 in.

Vertical bulkheads extending the entire height of the wall should be provided at about 26 ft. on center to control the flow of the grout horizontally. In a hollow-unit masonry wall, such barriers are made by placing mortar on cross webs and blocking the bond-beam units with masonry bats set in mortar. In a multi-wythe wall, the barriers are laid into the grout spaces as the wall is erected. In addition to confining grout to a manageable area, these barriers may be used as stiffeners or points to locate wall bracing.

Proper preparation of grout spaces is one of the most important features of high-lift grouting. It is necessary before grouting to remove all mortar droppings and debris through cleanout openings. Not less than 3 in. in size, a cleanout opening is located at the bottom of every core in hollow-unit reinforced masonry containing dowels or vertical reinforcement, and in at least every other core that is grouted but has no vertical bars. Cleanouts should be provided in solidly grouted masonry at a maximum spacing of 32 in. on center. In a two-wythe masonry wall, the cleanouts are provided at the bottom of the wall by omitting alternate units in the first course of one wythe. The governing standard or building code should be consulted to verify requirements for cleanout openings.

Cleanout openings in the face shells of units should be made before the units are laid. A special scored unit that permits easy removal of part of a face shell is occasionally used. Also, an alternate cleanout design makes use of header units as shown in Fig. 6-61.

It is considered good practice to cover the bottom of a grout space with a 2- to 3-in. layer of sand or a polyethylene sheet to act as a bond-breaker for the mortar droppings. The grout space is flushed at least twice a day (at midday and quitting time) with a high-pressure stream of water or, to keep the masonry from being moistened unnecessarily, the mortar droppings and projections are dislodged with a long pole or rod as the work progresses. After the masonry units are laid, the sand or polyethylene sheet is removed; compressed air is used to blow any remaining mortar out of the grout space; and the space is checked for cleanliness and the reinforcement for position. A mirror is a good inspection tool for looking up into the grout space through a cleanout opening.

Before grouting, the cleanout openings are closed by inserting masonry units or the face shells that were left out, or by placing formwork over the openings to allow grouting right up to the wall face. Grouting need not be delayed until the face-shell plugs or cleanout closure units are cured, but they should be adequately braced to resist the grout pressure.

In high-lift grouting, intermediate horizontal construction joints are usually not permitted. Once the grouting of a wall section is started, one pour* of grout to the top of the wall (in 5-ft. maximum lifts) should be planned for a workday. Should a blowout, an equipment breakdown, or any other emergency stop the grouting operation, a construction joint may be used if

approved by the inspector. The alternatives are to wash out the fresh grout or else rebuild the wall.

For economical placement a uniform 5-ft. lift of grout is generally pumped into place and immediately (not more than 10 minutes later) consolidated by vibration. Each succeeding lift of grout is pumped and consolidated after an appropriate lapse of time, a minimum of 30 and a maximum of 60 minutes (depending upon weather conditions and masonry absorption rates); this allows time for settlement shrinkage and the absorption of excess water by the masonry units. This waiting period also reduces the hydrostatic pressure of the grout and thus the possibility of a blowout. In each lift, the top 12 to 18 in. of grout is reconsolidated before or during placement of the succeeding lift.

In multi-wythe construction, the total length of wall that can be grouted in one pour is limited. It is determined by the number of sections (bounded by vertical bulkheads) that can be grouted to maintain the one-hour maximum interval between successive lifts in any section.

The maximum height of a pour is limited by Table 2-8 and by practical considerations, such as segregation of grout, the effect of dry grout deposits left on the masonry units and the reinforcing steel, and the ability to consolidate the grout effectively. The height of pour may be governed by story height, and thus 8-in., single-wythe walls may have a 20-ft. height of pour. When the grout pour exceeds 8 ft. in height, building codes sometimes require special inspection of the work.

Extreme care should be used to prevent grout from staining any masonry wall that will be exposed to view, especially in architectural uses of concrete masonry. If grout does contact the face of the masonry, it should be removed immediately. Also, soon after the wall has been fully grouted, all exposed faces showing grout scum or stains (percolated through the masonry and joints) should be washed down thoroughly with a high-pressure stream of water. If necessary, further cleaning may be done after curing and before final acceptance by the architect.

The time- and money-saving advantages of high-lift grouting on large projects are obvious. The vertical steel can be placed after the wall is erected to a full story height, and even on a job of moderate size the grout can be supplied by a ready mixed concrete producer and pumped in a continuous operation. The main disadvantages of high-lift grouting may be the need for a grout pump or other means of pouring grout rapidly, and the requirement for cleanout openings at the base of the wall.

Tooling Mortar Joints

Weathertight joints and the neat appearance of concrete masonry walls are dependent on proper tooling;

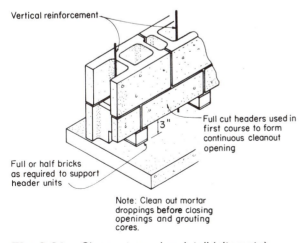

Vertical reinforcement

3"

Full cut headers used in first course to form continuous cleanout opening

Full or half bricks as required to support header units

Note: Clean out mortar droppings **before** closing openings and grouting cores.

Fig. 6-61. Cleanout opening detail (alternate).

*See footnote in section on "Low-Lift Grouting."

that is, compressing and shaping the mortar face of a joint with a special tool slightly larger than the joint. (Fig. 6-62). After a wall section has been laid and the mortar has become thumbprint-hard (when a clear thumbprint can be impressed and the cement paste does not adhere to the thumb), the mortar joints are usually considered ready for tooling. Sometimes units are laid almost up to quitting time and the last units are tooled early. The early tooling brings more paste to the surface and the joints are lighter in color. If all joints are tooled early, then all the mortar surfaces are more uniform than mortar tooled at varying time periods after placement. During hardening, mortar has a tendency to shrink slightly and pull away from the edges of the masonry units. Proper use of a jointing tool restores the intimate contact between the mortar and the units and helps to make joints weathertight. Proper tooling also produces uniform joints with sharp, clean lines.

Horizontal joints should be tooled before vertical joints. A jointer for tooling horizontal joints should be at least 22 in. long, preferably longer, and upturned on at least one end to prevent gouging (Fig. 6-62), while a jointer for vertical joints is small and S-shaped (Fig. 6-63). Vertical joints should be struck upward to close the gap between extruded mortar and the unit. Plexiglass jointers are available to avoid staining white or light-colored mortar joints. After the joints have been tooled, any mortar burrs should be trimmed off flush with the face of the wall by using a trowel (Fig. 6-64) and then dressed with a burlap bag, carpet, or brush (Fig. 6-65).

The principal types of mortar joints used in concrete masonry are shown in Fig. 6-66. The concave, vee, raked, and beaded types need special jointing tools, whereas the flush, struck, and weathered types are finished with a trowel. The extruded (also called skintled or weeping) type is made by using extra mortar so it can be squeezed out or extruded as the units are laid; it is not trimmed off but left to harden. Extruded joints are not recommended for walls subject to heavy rains, high winds, freezing temperatures, or those to be painted.

Vee joints are usually narrow in appearance and have sharp shadow lines. Concave joints (pictured in Fig. 6-62) have less pronounced shadows. Both types, when properly tooled, are very effective in resisting rain penetration. They are recommended for exterior weathertight walls, as is the weathered type.

Flush joints are simple to make because the excess mortar is simply trimmed off with a trowel (striking upward rather than downward) and the flush face rubbed with a carpet-covered wood float. The mortar is not compacted by tooling, and small hairline cracks produced when the mortar is pulled away from the units by the trimming action may permit infiltration of water into the wall. Flush joints are used in walls that will be plastered.

Fig. 6-62. Tooling a horizontal joint (concave type).

Fig. 6-63. Tooling a vertical joint.

Raked joints (pictured in Fig. 6-67) are made with a special tool—a joint raker or skate—to remove the mortar to a certain depth, which should not be more than ½ in. These joints produce dark shadows that accent the masonry pattern. Since their ledges hold rain, snow, or ice that may affect the watertightness of the wall, they are best suited to dry climates or interior use. Raked joints also provide less corrosion protection for joint reinforcement. Without special consideration,

Fig. 6-64. Mortar burrs are trimmed off after the joints are tooled.

Fig. 6-65. Dressing masonry with a brush after tooling.

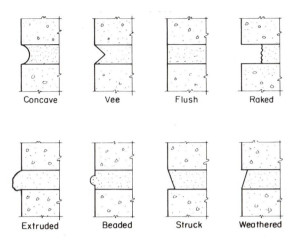

Concave Vee Flush Raked

Extruded Beaded Struck Weathered

Fig. 6-66. Principal types of mortar joints.

very little mortar (⅛ in.) is left to protect the reinforcement from weather. Reinforcement should have a minimum mortar cover of ⅝ in. for masonry exposed to weather or earth.

Beaded joints are basically extruded joints that are tooled with a special bead-forming jointer. The beads protruding on the wall surface present strong shadow lines, but special care is required to obtain a straight line appearance.

The struck and weathered types of joints are generally used to emphasize horizontal lines. Struck joints are easy to make with a trowel, especially if the mason works from the inside of the wall. However, their small ledges do not shed water readily, making them unsuitable for use in areas where heavy rains, driving winds, or freezing temperatures are likely to occur. On the other hand, weathered joints—one of the types recommended for weathertight walls—shed water easily but require careful finishing; they must be worked with a trowel from below.

The overall appearance of a masonry wall depends not only on the joint treatment but also on the color uniformity of the joints. Although mortar shade is influenced to some degree by the moisture condition of the units and by the atmospheric conditions, it depends mainly on the uniformity of the mortar mix and the time of joint tooling. The amount of water used in mixing colored mortar greatly influences the shade and thus requires accurate control. Colored mortar is usually not retempered as additional water may cause a significant lightening of the mortar. Any mortar that has become too stiff for use should be discarded. A darker color results if the mortar is tooled when relatively hard rather than reasonably plastic, but some masons consider mortar ready for tooling only when thumbprint-hard. Uniform time of tooling is important for obtaining uniformly colored joints.

Fig. 6-67. Raked joints accent a split-block wall.

Patching and Pointing

In spite of good workmanship, joint patching or pointing (the act of troweling additional mortar into a joint shortly after masonry units are laid) may sometimes be necessary. Mortar in a head joint may have fallen out while the units were being placed, a mortar crack may have formed while the units were being aligned, or units may have been dislodged by other construction activity. Furthermore, there may not have been enough mortar in a joint to fill the space left by a broken corner or edge. Sufficient additional mortar should be forced into such spots to completely fill the joints.

Patching or pointing is done preferably while the mortar in the joint is still fresh and plastic. If the back of the face shell can be reached when forcing additional mortar into the joint, the mason provides a backstop, such as the handle of a hammer. Any depressions and holes made by nails or line pins are filled with fresh mortar before tooling.

When patching must be done after the mortar has hardened, the joint is chiseled out to a depth of about ½ in., thoroughly wetted, and tuckpointed with fresh mortar. The replacement of hardened mortar by fresh mortar in cutout or defective joints is usually called tuckpointing (see Chapter 9). The term pointing usually refers to the repair of fresh joints.

Keeping Out Termites

Termites can squeeze through openings as small as ¹/₃₂ of an inch and their shelter tubes have been found in the cavities of concrete masonry. Thus, concrete masonry requires the following special precautions beyond those usually made in construction that needs control and protection against these wood-eating insects.

Masonry foundations, piers, and basement walls should be treated with a chemical such as aldrin, dieldrin, or chlordane. The chemical—at least 2 gal. for each 10 lin.ft.—is injected or buried below the surface of the ground along the wall or the base.

In addition, all masonry foundation walls and piers should be capped with cast-in-place concrete that is at least 4 in. thick and reinforced with two No. 3 bars. The capping should extend the full width of the wall and across the voids in veneered or cavity walls. The top of the cap should be at least 8 in. above grade. Solid masonry units are not acceptable as a substitute for cast-in-place capping for termite protection.

Bracing of Walls

Too frequently a freshly laid masonry wall will be blown over by wind. Such losses can be prevented. Good construction practice, most building codes, and the U.S. Occupational Safety and Health Administration (OSHA) regulations* require that new walls be braced for wind (Fig. 6-68).

Concrete masonry walls usually are not designed to be freestanding. Wind pressure can create four times as much bending stress in a new freestanding wall as in a finished wall of a building. This stress occurs at the bottom of the wall where flashing or lack of bond decreases the wall strength to resist tensile wind forces, and fresh mortar in a wall is weak. The bracing should be designed to resist wind pressure as required by local regulations for building design.

Bracing should be provided if the height of the wall exceeds that given in Fig. 6-69 for various peak wind velocities. Where bracing is used, the heights shown are the safe heights above the bracing. For example, to withstand wind gusts of 50 mph, the freestanding height (without bracing) of a 10-in.-thick wall weighing 67 psf or less should not exceed 7½ ft. The family of curves in the figure is based on the assumptions that the mortar has no tensile strength and the walls are freestanding, nonreinforced, ungrouted concrete masonry. For cavity walls the thickness should be assumed to be two-thirds the sum of the thicknesses of the two wythes.

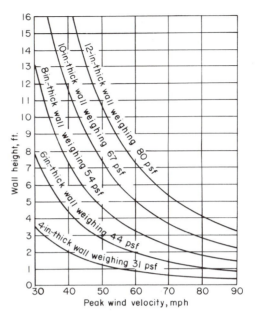

Fig. 6-69. Maximum unsupported heights of nonreinforced, ungrouted concrete masonry walls during construction. Source: Ref. 52.

Quality of Construction

Some time before, during, and after construction questions are asked about the quality of construction. This is investigated by quality control and quality assurance** personnel. It involves several factors:

1. The knowledge and attitude of those in charge determine how much quality is wanted and/or will be achieved.
2. The assignment of responsibility for quality construction depends upon the type of job.
3. The requirements for satisfactory quality in construction are well established but require familiarity.

The quality assurance program assures the owner that the building is constructed in accordance with the specifications and applicable standards. Quality assurance addresses the responsibilities of those involved with the project, materials control, inspection, laboratory and field testing, documentation of conditions or

Fig. 6-68. Newly laid concrete masonry walls should be braced against wind because of their low strength when freestanding.

---

*Sec. 1926.700(a) of the OSHA Construction Safety and Health Regulations adopts, by reference, Article 12.5 of the American National Standard Safety Requirements for Concrete Construction and Masonry Work (ANSI A10.9), which states: "Masonry walls shall be temporarily shored and braced until the designed lateral strength is reached, to prevent collapse due to wind or other forces."

**"Quality control" is the contractor's or manufacturer's effort to achieve a specified result. "Quality assurance" is the owner's effort to require a level of quality and determine its acceptability.

Fig. 6-70. Construction of a high-rise hotel with engineered load-bearing concrete masonry. Strict quality control is required.

The quality control inspector must also thoroughly understand the masonry specifications; reflecting the experience of the entire industry, each paragraph and sentence is packed with significance. Reference specifications are equally important, and copies should be readily available to the inspector. Moreover, since a competent inspector must know or be able to quickly determine many things, it is suggested that inspectors become familiar with this entire handbook. One of its purposes is to explain and set forth recommended practices for quality construction.

Quality and safety of construction can usually be equated, particularly in engineered masonry construction. In this type of work the importance of inspection is so great that allowable stresses should be reduced if there is no engineering or architectural supervision of the construction. ACI 121R, 530, 530.1, and the commentaries to these documents provide guidance and requirements for quality assurance and quality control.

Sampling and Testing Units

Compressive strength, absorption, weight, moisture content, and dimensions are tested on masonry units selected at the place of manufacture from the lots ready for delivery. At least 10 days should be allowed for the completion of the tests. Tests for drying shrinkage, however, must be conducted long before delivery. The allowable moisture content depends on the average annual relative humidity of the locale; U.S. data are given in Fig. 6-71.

Compression tests require testing machines of large capacity. The equipment in most commercial testing laboratories has a maximum capacity of 300,000 lb., while the ultimate strength of a single high-strength masonry unit may greatly exceed this. Therefore, ASTM C140 permits a unit to be "sawed into segments" if a lower-capacity testing machine has to be used. For example, a two-core unit could be cut down to a one-core unit.*

Making a Sample Panel

In concrete masonry work involving an architect or engineer, a sample panel of the wall built by the contractor at the start of the job is highly recommended. It is the responsibility of the quality assurance representative to approve the sample panel; however, the architect or owner should approve the overall appearance of the panel for important architectural work.

Often 64 in. long and 48 in. high, the sample panel should include the masonry reinforcement as well as show the workmanship, coursing, bonding, thickness,

materials meeting or not meeting the specifications, the lines of communication, and reporting. The contractor's or producer's quality control program is the production tool, or the actual construction, testing, and inspection, performed to demonstrate specification compliance.

For the owner and his quality assurance representative the key to success—a quality concrete masonry structure—is a quality conscious masonry contractor. The local masonry association can usually suggest a contractor or answer questions concerning contractors. Another good practice is to check the contractor's references and previous work.

In the field the roles of the quality control and quality assurance inspectors are much greater than overseeing a few tests and inspecting the completed structure. They must become very familiar with the job specifications and drawings. As this is a lengthy process, it is a great advantage to have the engineer or architect or their authorized representatives on the jobsite.

*See Ref. 44.

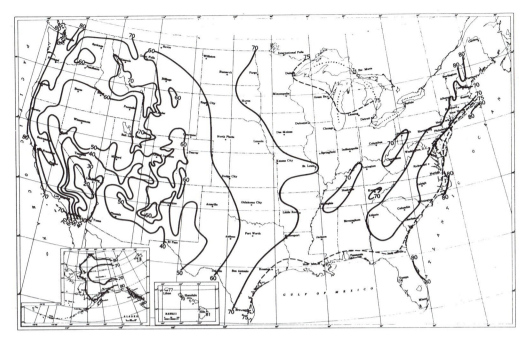

Fig. 6-71. Mean (average) annual realtive humidities, percent, for the United States. Prepared by Environmental Science Services Administration, U.S. Department of Commerce, 1968.

tooling of joints, range of unit texture and color, and mortar color—all as specified or selected. Of course, the finished work should match the sample panel.

The sample panel may or may not be a part of the building. If a part of the building, it should be located such that access to it for approval/rejection of future work can be established. If the sample panel is a separate panel and not part of the building, it should also be used for acceptance/rejection of such things as cleaning of masonry and adequacy of surface treatments. The sample panel should not be removed until the owner accepts the building.

Testing Project Mortar

Before construction begins, 2-in. mortar test cubes may be prepared in the laboratory (in accordance with ASTM C270) to check the compatibility and compressive strength characteristics of the cementitious materials and sand. These cube tests are made only if the mortar mixture has been specified by strength, not proportions. Because a much greater water content is used during construction, the strengths of any cubes made from mortar sampled at the project cannot be compared to the ASTM C270 strength specifications. However, periodic cube tests made from field mortar can be used to check the uniformity of batching operations. They will be even more useful if they are referenced to accurately measured batches.

ASTM specification C780 is currently useful for recommending specific masonry mortar field tests and for interpreting the test results. These test results provide a more reliable measure of uniformity of field batching than if periodic cube tests were made according to ASTM C270.

Testing Masonry Prisms and Grout

Prism strength tests (Fig. 6-72), discussed in Chapter 3, may be required not only in advance of design but also as part of quality control in the field, especially on projects involving engineered concrete masonry. During construction, one test is usually required for every 5,000 sq.ft. of wall; each test includes three prisms and not less than three prisms are made per story. By the probabilities laws, some test values may fall below the specified prism strength and this should be evaluated in light of the discussion in Chapter 3.

Separate field tests for compressive strength of grout, discussed in Chapter 2, may be another requirement for quality control. ACI 530.1 requires grout to have a minimum compressive strength of 2,000 psi at 28 days when tested according to ASTM C1019, unless grout is specified according to ASTM C476 proportions.

Destructive Testing of Grouted Walls

Some quality control practices for grouted concrete masonry walls involve the use of destructive testing;

Fig. 6-72. Method of testing the compressive strength of a masonry prism.

that is, removal of cores drilled through the completed walls and submittal of these cores to visual inspection and physical tests. This practice not only causes damage to the walls but also is expensive, is not directly repeatable, and does not necessarily determine locations of imperfections.

Where grouted walls are to be cored, the cores should be taken at least 6 ft. from a corner and not under an opening in the wall. The government code or specification should be checked for the size of core required, and at least two cores should be tested for each project. Those codes that require core tests usually specify a minimum compressive strength of 750 psi at 7 days and 1,500 psi at 28 days.

It should be recognized that, while the cores are drilled laterally through the wall, they must be test-loaded along the core axis, that is, at right angles to the axis of the applied loads on the wall. Therefore, it is unlikely that the test result is a true measure of wall strength. A truer representation of the strength of the masonry wall in the direction of the anticipated applied loads is obtained by testing a sample prism cut out of the wall.

Generally through-the-wall cores are taken to check their compressive strength. However, they may also be taken to check the shear bond strength between the masonry units and grout. The test determines the unit force required to shear the masonry face shells from the grout core for each face. Some building codes specify a minimum shear bond strength of 100 psi at 28 days.

Nondestructive Testing of Grouted Walls

Without damage to the walls, two tests permit considerable wall area to be inspected for major variations or flaws. These tests consist of ultrasonic and radiographic inspection.

The ultrasonic test measures the time that sound waves take to pass through the thickness of a grouted wall. When the time interval indicating good grouting is known, the time measure can be used as a quality control tool. A relatively long time interval indicates the presence of flaws such as voids, cracks, or honeycomb. The moisture content of the wall affects the time interval but does not interfere with identifying variations within the wall. Although ultrasonic inspection could cost more than drilling and testing cores, the more complete coverage available may well justify the increased cost.

In the radiographic test, a beam of either X-ray or cobalt 60 isotope radiation is passed through a grouted wall to expose a photographic film on the reverse side. The film shows variations in wall density as gradations of black and white, and thus the sensitivity of the procedure depends to a large extent on the type of wall constructed. Since radiation is involved, a hazard does exist with this method of inspection.

Dry-Stacked, Surface-Bonded Masonry

Concrete masonry walls dry-stacked without mortar in the bed and head joints and to which a layer of glass-fiber reinforced cement paste or mortar is applied provides an alternate concrete masonry construction method. The technique is referred to as dry-stacked, surface-bonded masonry (Fig. 6-73).

Manufacturers provide proprietary mixtures consisting of cement and alkali-resistant glass fibers; the mixtures also may contain fine aggregate. These formulations are standardized by product specification, ASTM C 887. Guidance as to construction practices are provided in ASTM C 946.

Masonry walls are built using a full mortar bed under the first course of block laid on the foundation. Units abut each other without any mortar in head joints. Subsequent courses of block are dry-stacked, again without benefit of mortar in the bed or head joints. Units are laid in a running bond pattern. Wedges or shims assist in maintaining level courses. Throughout the height of a wall, leveling courses bedded in mortar are laid when vertical differences exceed ½ in. in 10 feet for a single course. Leveling courses are also installed at floor levels.

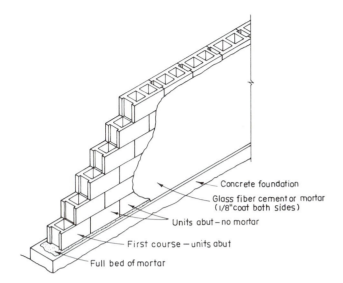

Concrete foundation

Glass fiber cement or mortar (1/8"coat both sides)

Units abut — no mortar

First course — units abut

Full bed of mortar

Fig. 6-73. Dry-stacked, surface-bonded masonry.

After dry-stacking, the surface bonding mixture is combined with water and hand-or machine-applied to both sides of the wall, providing a coating at least ⅛ in. thick. Spray or trowel finishes after spray or hand applications are attainable.

A wide variety of applied finishes are possible with concrete masonry construction. The finish to use in any particular case will be governed by the type of structure in which the walls will be used, the climatic conditions to which they will be exposed, and the architectural effects desired. Some popular finishes are described in this chapter.

Paints

The main purposes of painting concrete masonry walls are to add a fresh appearance and color and to alter the surface texture and pattern. Additional purposes are to reduce the passage of sound through the wall and, as discussed below, to bar the passage of moisture. The paint selected should not only achieve these goals but also attach itself tenaciously to the underlying surface, retain its appearance for a long time, and be economical.

Paints are pigmented coatings that form opaque films. Paint mixtures have minute solid particles of pigment suspended in liquid referred to as the vehicle. The pigment provides hiding power (conceals the surface beneath) and color. The vehicle includes two elements: the volatile solvent or thinner that supplies the desired consistency for application, and the binder that bonds the pigment particles into a cohesive paint film during the drying and hardening process. Some paints dry and harden by evaporation of the solvent; some, by oxidation. Most of the other types involve chemical reactions as well as evaporation of the solvent.

Some paints breathe; that is, they are permeable and allow passage of water vapor. The others are non-breathing or impermeable. Impermeable paints should be applied to the side of a wall where moisture enters; permeable paints, to the surface where moisture exists. If the surface from which moisture is attempting to leave is coated with an impermeable paint, blistering will occur and the paint will peel.

There are several other items that must be considered when selecting paint for a concrete masonry surface. Among them are: (1) whether or not the paint will be damaged by the probable presence of alkalies, and (2) whether the surface texture requires alteration before application of the paint.

Types Commonly Used

Many products have been marketed for use in painting concrete masonry walls and they have had varying degrees of success. The basic constituents and pertinent characteristics of paints now commonly used on concrete masonry walls are described below. Urethanes, polyesters, and epoxies are also used successfully.

Portland Cement Paints

Of the various types of paint used on concrete masonry construction, those with a portland cement base (Fig. 7-1) have the longest service record. U.S. Federal Specification TT-P-21, which gives requirements for composition, provides for two types and two classes of portland cement paint: Type I (containing a minimum of 65% portland cement by weight) is for general use; Type II (with at least 80% portland cement by weight) is for maximum durability. Under each type, Class A contains no aggregate filler and is for general use, while Class B contains 20 to 40% siliceous sand filler for use on open-textured surfaces.

A concrete masonry surface should be damp at the time of application of a portland cement paint (Fig. 7-2). The setting and curing require the presence of water, a favorable temperature, and sufficient time for hydration. If the paint is modified with latex, moist curing is not necessary because the latex retains sufficient moisture in the paint film for hydration.

Fig. 7-1. A stippled, heavy cement-based coating on a concrete masonry screen wall.

Fig. 7-2. The concrete masonry surface is uniformly dampened with a water spray just before application of a portland cement paint. A garden pressure sprayer with a fine-fog nozzle is recommended.

Although portland cement paints may be made on the job, the best results (uniform color, durability) are most often obtained by using products marketed in prepared form. Porland cement paints form hard, flat, porous films that readily permit passage of water vapor. These paints are not harmed by the presence of alkalies and may be applied to freshly erected concrete masonry surfaces. However, the results will be better if painting is deferred at least three weeks.

Latex Paints

Latex paints are water-thinned; that is, they are based on aqueous emulsions of various resinous materials such as acrylic resin, polyvinyl acetate, and styrene butadiene. With an ever increasing use of polymers, blends, and modifications of the base resins, the distinction between these materials has been blurred to the point where any such classification is somewhat meaningless. They all dry very rapidly by evaporation of water, followed by coalescence of the resin particles.

Careful surface preparation is required for latex paints since they do not adhere readily to chalked, dirty, or glossy surfaces. However, these paints are easy to apply and have little odor. Also, they are economical, nonflammable, breathing paints that are not damaged by alkalies. They have excellent color retention and are very durable in normal environments. Those containing acrylics are more expensive than the other latex types, but field experience has shown them to give the best performance.

Oil-Bases Paints and Oil-Alkyds

Oil-based paints contain drying oils as the binder and are nonbreathing. They are similar to conventional house paints. Easy to use, these paints are durable under some exposures but not particularly hard or resistant to abrasion, chemicals, or strong solvents. They are also damaged in the presence of alkalies.

Oil-based paints are often modified with alkyd resins to improve resistance to alkalies, reduce drying time, and improve performance in other ways. When the substitution of resins for oil is high, they are referred to as varnish-based paints. The oil-alkyds and even the varnish-based paints may be susceptible to damage from alkalies. Oil-alkyds also are nonbreathing paints.

There are few instances where serious consideration should be given to an oil-based or oil-alkyd paint for use on concrete masonry.

Rubber-Based Paints

Formulated with a chlorinated natural rubber or a styrene-butadiene resin, these paints form a nonbreathing film with alkali and acid resistance. They are used not only for exterior masonry surfaces but also for interior ones that are wet, humid, or subject to frequent washing (swimming pools, wash and shower rooms, kitchens, and laundries); that is, where alkali resistance is important and where requirements for resistance to the entrance of moisture are greater than can be supplied by latex paints. Rubber-based paints may be used as primers under less resistant paints.

Surface Preparation

Regardless of the type of paint selected, its success or failure can be dependent upon the adequacy of surface preparation. In most cases success is more assured if the concrete masonry surface has aged at least six months before painting; this is due to the dampness usually present in a new masonry wall and to alkalinity of the surface. If the paint is not sensitive to either moisture or alkalies (such as a portland cement paint or latex paint), a long aging period is unnecessary. Earlier use of oil- or alkyd-base paint, which is subject to damage from alkalinity, is possible if the surface is neutralized by pretreatment with a 3% solution of phosphoric acid followed by a 2% solution of zinc chloride. However, this procedure is now rare because of the success of other paints not requiring pretreatment, such as portland cement and latex paints.

For paint to adhere, concrete masonry surfaces must be free of dirt, dust, grease, oil, and efflorescence. Dirt and dust may be removed by air-blowing, brushing, scrubbing, or hosing. If a surface is extremely dirty, wet or dry sandblasting, waterblasting, or steam-cleaning may be used. Grease and oil are removed by applying a 10% solution of caustic soda, trisodium phosphate, or detergents specially formulated for use on concrete. Efflorescence is cleaned off by brushing or light sandblasting. After any of these treatments, the surface should be thoroughly flushed with clean water.

Fill coats, also called fillers or primer-sealers, are sometimes used to fill voids in open or coarse-textured concrete masonry surfaces. Applied by brush before the finish coat(s), the fill coats usually contain white portland cement and fine siliceous sand. If acrylic latex or polyvinyl acetate latex is included in the mixture, moist curing is not required. Fill coats impair sound absorption but improve the sound transmission loss of a concrete masonry wall.

For more information on preparation for surface coatings, see ASTM D4261, Standard Practice for Surface Cleaning Concrete Unit Masonry for Coating and "Cleaning Concrete Masonry Surfaces," Chapter 9.

Paint Preparation and Application

Paint must be thoroughly stirred just prior to application. Power stirrers and automatic shakers are becoming more common, but they are not recommended for latex paints because of the possibility of foaming. Hand stirrers should have a broad, flat paddle.

Fig. 7-3. Typical brushes used in applying portland cement paint: *(left to right)* ordinary scrub brush, window brush, brush with detachable handle, and fender brush.

Fig. 7-4. Portland cement paint is applied at the joints first.

Thinning of paint should only be done in accordance with the manufacturer's directions; excessive paint thinning will result in coatings of low durability. Color tinting should also be done carefully in accordance with the manufacturer's suggestions.

The paints commonly used on concrete masonry are applied by brush, roller, or spray as described below. Roller application is often used for large areas.

1. *Portland cement paints* are applied to damp surfaces by brush with bristles no more than 2 in. in length (Fig. 7-3). The paint should be *scrubbed* into the surface as shown in Fig. 7-4. An interval of at least 12 hours should be allowed between coats. After completion of the final coat, 48-hour

moist curing is necessary if the paint is not modified with latex.

2. *Latex paints* may be applied to dry or damp surfaces by roller or spray, but preferably by long-fiber, tapered nylon brush 4 to 6 in. wide (soaked in water for two hours before use). When the surface is moderately porous or extremely dry weather prevails, it is advisable to dampen the surface. These paints dry throughout as soon as the water of emulsion has evaporated, usually in 30 to 60 mintues, and require no moist curing.

3. *Oil-based and oil-alkyd paints* should not be applied during damp or humid weather or when the temperature is below 50 deg. F. Pretreatment surface neutralization is recommended if the masonry is less than 6 months old (see "Surface Preparation"). Application is by brush, roller, or spray (usually by brush) to a dry surface. Each coat should be allowed to dry at least 24 hours and preferably 48 hours before application of succeeding coats.

4. *Rubber-based paints* are usually applied by brush to dry or slightly damp surfaces. Two or three coats are necessary to achieve adequate film thickness, and the first coat is usually thinned in accordance with the manufacturer's recommendations. A 48-hour delay is recommended between coats. Recoating should be performed with care. Because of the strong solvents used in rubber-based paints, a second coat tends to lift or attack the original coat.

Caution: Most paints are flammable, and some have solvents that are highly flammable. Therefore, adequate ventilation during painting should be provided according to manufacturers' recommendations.

Clear Coatings

In many areas, architects specify application of a clear coating or water repellent on concrete masonry structures. This is done to render the surface water-repellent and thus protect the masonry from soiling and surface attack by air-borne pollutants, as well as to facilitate cleaning. Further advantages are to prevent the surface from darkening when wet and to accentuate surface color. In some areas, where weather exposure is not severe and air pollution is low, coatings may not be necessary.

The coating selected should be water-clear and capable of being absorbed into the surface. It should also be long-lasting and not subject to discoloration with time and exposure. Good service has been obtained with coatings based on a methyl methacrylate form of acrylic resin,* silanes, siloxanes, and other compounds. Brush or spray application of one or two coats of a relatively low-solid-content coating is usually satisfactory. Consult the coating manufacturer as to the life

expectancy of the coating or sealer. Building owners should be prepared to reapply certain coating or sealers every 5 to 10 years.

Stains

Decorative staining of concrete masonry walls can give good results if the proper stain is selected and then applied correctly. However, there may be some drawbacks to staining. For example, color applied after concrete hardens is not as long-lasting as that incorporated into the concrete mix during block manufacture. Also, the color effects or shades of a stain may vary.

Types

Several types of stains may be used on concrete masonry. They include:

1. *Oil stains,* which sometimes require aging of the wall for several months before application of stain, or pretreatment of the surface to inhibit reaction between alkalies and the oil vehicle in the stain. Many stains suitable for wood are suitable for concrete masonry.

2. *Metallic salt stains,* which are slightly acid solutions of salts that result in the deposit of colored metallic oxide or hydroxide in the surface pores. These deposits are not soluble in water.

3. *Organic dyes,* which contain analine dyes.

Application

For best results, staining should be delayed at least 30 to 45 days after the concrete masonry structure is built. The surface must be dry and clean—free of oil, grease, paint, and wax. Acid etching is usually not necessary or advisable because of the porosity of a concrete masonry surface. Also, two or more stain applications may be required to secure the depth of color desired.

Each coat of stain should throughly saturate the surface and be evenly applied by a constant number of passes of a brush, roller, or low-pressure spray. Care should be excercised so that the stain does not streak or overlap into a dried area. From four to five days should elapse between coats, depending upon the masonry surface, ambient conditions, and the stain used. It will often take three or four days for a stained surface to reach its final color.

---

*See Ref. 45.

Portland Cement Plaster Finishes

Portland cement plaster and portland cement stucco are the same finishing material. It can be applied to the surface of any concrete masonry structure. Although stucco is the term often associated with exterior use of the material, the term plaster denotes interior or exterior use (although in some areas plaster denotes interior use only). Portland cement plaster is a combination of portland cement, masonry cement, or plastic cement with sand, water, and perhaps a plasticizing agent such as lime. A color pigment may be used in the finish coat.

Porland cement plaster has many of the desirable properties of concrete and, when properly applied, forms a durable, hard, strong, and decorative finish. There is an unlimited variety of textures, colors, and patterns possible. The rougher textures help conceal slight color variations, lap joints, uneven dirt accumulation, and streaking.

Portland cement plaster finishes are primarily used for exterior walls, but they are also particularly well suited for interior high-moisture locations such as kitchens, laundries, saunas, bathrooms, and various industrial facilities. The use of exterior plaster (stucco) involves considerations of water penetration, corrosion of reinforcement and accessories, and stresses due to wider variations of temperature and humidity than normally present in interiors.

Plaster should conform to ASTM C926, Standard Specification for Application of Portland Cement-Based Plaster. Also, refer to ASTM Standard Specifications C631, C932, C933, and C1032. Furthermore, a full discussion of portland cement plastering apears in the PCA *Plasterer's Manual.** A brief review of portland cement plaster finishes follows.

Jointing for Crack Control

The proper design and selection of materials for a concrete masonry wall can substantially minimize or eliminate undesirable cracking of its portland cement plaster finish. Cracks can develop in this finish through many causes or combinations of causes; for example, shrinkage stresses; building movements; foundation settlement; restraint from lighting and plumbing fixtures, intersecting walls or ceilings, pilasters, and corners; weak sections due to cross-section changes such as at openings; and construction joints.

It is difficult to anticipate or prevent cracks from all these possible causes, but they can be largely controlled by means of control joints (Fig. 7-5). Joints should be installed directly over any previous joints in the base. For plaster bonded directly onto concrete masonry, additional joints are not usually needed.

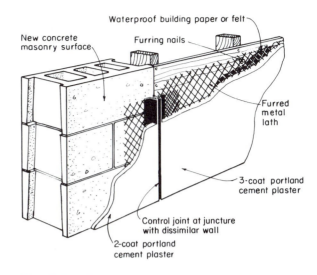

Fig. 7-5. Control joint at the juncture of dissimilar walls: *(left)* two-coat plaster applied directly to concrete masonry; *(right)* three-coat plaster on metal lath over wood construction.

Concrete masonry walls and ceilings that use metal lath for the plaster base should be divided into square or rectangular panels not exceeding 140 to 150 sq. ft. A control joint should be placed at least every 10 to 12 ft. Ideally the panels should have a length to width ratio of 1 to 1, but in no case should it exceed 2½ to 1. The metal reinforcement in the plaster must not extend across control joints, and the material used for the control joints on exterior surfaces should be weathertight and corrosion-resistant.

Mixes

Data for plaster mixes are given in Tables 7-1, 7-2, and 7-3. Note that lime should not be added when masonry cement or plastic cement is used. These cements already contain plasticizers and only sand and water need be added, thus simplifying jobsite proportioning and mixing.

When proportioned in accordance with the requirements of Table 7-2, a good plastering mix will be recognized by its workability, ease of troweling, adhesion to bases, and ability to attach itself to surfaces without sagging. Batch-to-batch uniformity will help assure uniform suction for subsequent coats and color uniformity.

Uniform measuring and batching methods are important. All ingredients should be thoroughly mixed (preferably in a power mixer) with the amount of water needed to produce a plaster of workable consistency.

---

*See Ref. 18.

Table 7-1. Permissible Mixes for Portland Cement Plaster Base Coats*

Type of plaster base	Plaster mix symbols	
	First coat (scratch coat)	Second coat (brown coat)
Concrete masonry**	L M P	L M P
Metal reinforcement†	C L CM M CP	C, L, M or CM L CM or M M CP or P

*Adapted from ASTM C926.
**High-absorption bases such as concrete masonry should be moistened prior to scratch coat application.
†Metal reinforcement with paper backing may require dampening of paper prior to application of plaster.

Natural or synthetic fibers are sometimes added to plaster to improve cracking resistance and pumpability. The fibers should not be used in excess of 2 lb. per cu. ft. of cementitious material to avoid workability and mixing problems.

Surface Preparation

Concrete masonry provides an excellent base for plaster because of its rigidity and excellent bonding characteristics. Bond occurs both mechanically and chemically. Mechanically, bond results from keying; that is, interlocking of plaster with the open texture in the concrete masonry surface. Suction by the masonry also improves mechanical bond; that is, paste is drawn into minute pores of the surface. Chemically, the similar materials adhere well to each other.

Table 7-2. Proportions for Portland Cement Plaster Base Coats*

Plaster mix symbols	Cementitious materials, parts by volume				Aggregate volume per sum of cementitious materials**		
	Portland cement	Lime**	Masonry cement†	Plastic cement	First coat	Second coat††	First and second coats
C	1	0 to ¾	—	—	2½ to 4	3 to 5	3 to 4
CM	1	—	1 to 2	—	2½ to 4	3 to 5	1½ to 2
L	1	¾ to 1½	—	—	2½ to 4	3 to 5	2 to 3
M	—	—	1	—	2½ to 4	3 to 5	—
CP	1	—	—	1	2½ to 4	3 to 5	2 to 3
P	—	—	—	1	2½ to 4	3 to 5	3 to 4

*Adapted from ASTM C926.
**Variations in lime, sand, or perlite contents are given due to variations in local sands, and the fact that higher lime content will permit use of higher sand content. The workability of the plaster mix will govern the amounts of lime and sand. To determine the volume (in parts) of aggregate to use, add up the parts of cementitious materials and multiply by a number within the range shown in the last three columns above.
†Type N masonry cement.
††Within the limits shown, the same or greater proportions of sand should be used in the second coat as in the first coat.

Mixing time should be a minimum of 2 minutes after all materials are in the mixer, or until the mix is uniform in color. The size of a batch should be that which can be used immediately or in no more than 2½ hours. Remixing, to restore plasticity with the addition of water, is permissible within the same time limits.

Color pigments are often used in the finish coat, which is usually a factory-prepared stucco* finish mix. It should be noted that factory-prepared finish mixes assure greater uniformity of color than job-preared mixes, and the manufacturer's recommendations should be closely followed. If the finish coat is job-mixed, truer color and a more pleasing appearance will be obtained when a white portland cement and a fine-graded, light-colored sand are used.

A new concrete masonry surface can be used as plaster base with minimum consideration for surface preparation. For best results, the concrete masonry units should have an open texture and be laid with struck joints (Fig. 7-6). The surface should be free of oil, dirt, or other materials that reduce bond; then prior to application of plaster, it should be uniformly dampened, but not saturated, with clean water using a fine-fog nozzle (Fig. 7-7).

An old concrete masonry surface should be assessed to establish its bonding characteristics. A surface having the desired texture and cleanliness will perform as

*In some areas the word stucco refers only to the finish coat while in most other areas it refers to the entire thickness.

Table 7-3. Proportions for Job-Mixed Portland Cement Plaster Finish Coat*

Plaster mix symbols	Cementitious materials, parts by volume				Aggregate volume per sum of cementitious materials††	
	Portland cement	Lime	Masonry cement†	Plastic cement	Sand	Perlite‡
F**	1	¾ to 1½	—	—	3	2
FL	1	1½ to 2	—	—	3	2½
FP**	—	—	—	1	1½	1
FM	—	—	1	—	1½	1½
FPM**	1	—	1	—	3	2

*Adapted from ASTM C926.
**Specify for surfaces subjected to abrasive treatment.
†Type N masonry cement.
††To obtain the volume (in parts) of aggregate to use, add up the parts of cementitious materials and multiply by the appropriate number shown in the last two columns above.
‡Use of perlite instead of sand is recommended only over basecoat plaster containing perlite aggregate and in areas not subject to impact.

Fig. 7-6. The natural rough texture of concrete block provides a good base for portland cement plaster. A suitable corner bead is provided by metal corner reinforcement with a series of coiled wires.

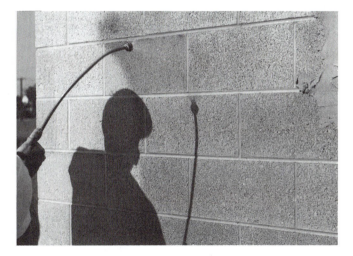

Fig. 7-7. A water spray is used to prepare the concrete block surface for its first coat of plaster.

well as a new masonry surface. If the masonry has been painted, sandblasting must be used to remove the paint and improve the bonding characteristics. Otherwise, furred metal lath* must be anchored to the wall over waterproof building paper or felt (Fig. 7-8).

The suitability of an unpainted concrete masonry surface as a plaster base can be tested by spraying it with clean water to see how quickly moisture is absorbed through suction. If water is readily absorbed, good suction is likely; if water droplets form and run down the surface, its suction is probably inadequate. For low-suction surfaces, bond must be increased by applying a bonding agent or a dash-bond coat (containing 1 part portland cement, 1 to 2 parts sand, and sufficient water for a thick paint-like consistency). In

lieu of a bonding agent or dash coat, bond must be provided by metal lath.

Application

Plaster may be applied by hand or machine in two or three coats in accordance with the required thicknesses given in Table 7-4. Horizontal (overhead) application

---

*Furring is a term applied to spacer elements used to maintain a gap between lath (or a finish such as wallboard) and the masonry. Lath is a material whose primary function is to serve as a plaster base; the most common types are expanded metal lath, expanded stucco mesh, hexagonal wire mesh (stucco netting), stucco mesh, and welded-wire fabric.

Fig. 7-8. Hand application of plaster on metal lath over waterproof building paper or felt. Dimples in the expanded metal lath provide proper furring or embedment of the metal.

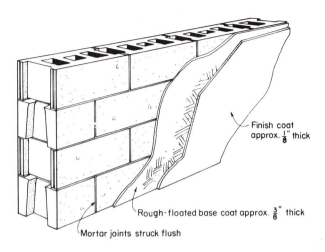

Finish coat approx. $\frac{1}{8}$" thick

Rough-floated base coat approx. $\frac{3}{8}$" thick

Mortar joints struck flush

Fig. 7-9. Portland cement plaster applied directly to concrete masonry.

Table 7-4. Thicknesses of Portland Cement Plaster*

Type of work		Type of plaster base		Coat thickness, in.							
				Coats on vertical surface				Coats on horizontal surface			
				1st	2nd	3rd**	Total	1st	2nd	3rd**	Total
Interior	3-coat work†	Metal reinforcement		$\frac{3}{8}$	$\frac{3}{8}$	$\frac{1}{8}$	$\frac{7}{8}$	$\frac{1}{4}$	$\frac{1}{4}$	$\frac{1}{8}$	$\frac{5}{8}$
		Solid base	Unit masonry	$\frac{1}{4}$	$\frac{1}{4}$	$\frac{1}{8}$	$\frac{5}{8}$	Use 2-coat work			
			Metal reinforcement over solid base	$\frac{1}{2}$	$\frac{1}{4}$	$\frac{1}{8}$	$\frac{7}{8}$	$\frac{1}{2}$	$\frac{1}{4}$	$\frac{1}{8}$	$\frac{7}{8}$
	2-coat work††	Solid base	Unit masonry	$\frac{3}{8}$	$\frac{1}{8}$	—	$\frac{1}{2}$	—	—	—	$\frac{3}{8}$ max.‡
Exterior	3-coat work†	Metal reinforcement		$\frac{3}{8}$	$\frac{3}{8}$	$\frac{1}{8}$	$\frac{7}{8}$	$\frac{1}{4}$	$\frac{1}{4}$	$\frac{1}{8}$	$\frac{5}{8}$
		Solid base	Unit masonry	$\frac{1}{4}$	$\frac{1}{4}$	$\frac{1}{8}$	$\frac{5}{8}$	Use 2-coat work			
			Metal reinforcement over solid base	$\frac{1}{2}$	$\frac{1}{4}$	$\frac{1}{8}$	$\frac{7}{8}$	$\frac{1}{2}$	$\frac{1}{4}$	$\frac{1}{8}$	$\frac{7}{8}$
	2-coat work††	Solid base	Unit masonry	$\frac{3}{8}$	$\frac{1}{8}$	—	$\frac{1}{2}$	—	—	—	$\frac{3}{8}$ max.‡

*Adapted from ASTM C926. Where a fire rating is required, plaster thickness should conform to the applicable building code or to an approved test assembly.

**The finish coat thickness may vary, provided that the total plaster thickness complies with this table and is sufficient to achieve the texture specified. For exposed-aggregate finishes, the second (brown) coat may become the "bedding" coat; it should be of sufficient thickness to receive and hold the aggregate specified.

†Where three-coat work is required, dash or brush coats of plaster are not acceptable as one of the three coats.

††For two-coat work, only the first and finish coats for vertical surfaces and the total plaster thickness for horizontal surfaces are indicated. The use of two coats is common practice when plaster is applied directly to vertical concrete masonry, and horizontal application seldom exceeds two coats.

‡On horizontal solid-base surfaces such as ceilings or soffits requiring more than $\frac{3}{8}$-in. plaster thickness to obtain a level plane, metal reinforcement should be attached to the concrete and the thickness specified for three-coat work on metal reinforcement over solid base applies. Where $\frac{3}{8}$-in. or less plaster thickness is required to level and decorate and there are no other requirements, a liquid bonding agent or dash-bond coat may be used.

seldom exceeds two coats, and two coats are often used when plaster is applied directly to concrete masonry as shown in Fig. 7-9. Three coats are applied when furred metal lath is used as a plaster base. A sprayed color coat—the finish coat—requires two applications, the first to ensure complete coverage and the second to obtain the desired texture.

Hand application (Fig. 7-8) involves traditional plastering tools and practices proven successful over many years. However, machine application (Fig. 7-10) is now widely used because it offers many advantages: pumping capability, speedier application, elimination of lap and joint marks, possibility of deeper and darker colors, and more uniform texture.

Machine application requires different procedures than those used for hand application. Basically, the mix is sprayed from the machine nozzle against the prepared base or previous coat. Nevertheless, if machines are to be used to best advantage, manufacturers' instructions should be carefully followed.

Proper consistency for machine-applied plaster is best determined by observing plaster during application. If the plaster is too fluid to build up the proper thickness, the water content is too high; if plaster will not pump or is exceedingly difficult to strike off, the water content is too low. Well-graded aggregate greatly improves the pumpability of plaster. To monitor consistency, specifications may require a slump test for plaster taken from the nozzle of the plastering hose.

Sufficient time, depending on ambient temperature, should be provided between coats to develop enough rigidity to resist cracking or other damage when the next coat is applied.

Fig. 7-10. Machine application of plaster.

Curing

Portland cement plaster requires moist curing after application in order to produce a strong, durable finish. Curing can be successfully accomplished by several methods, and the type and size of the structure and the climatic and job conditions will dictate the most suitable. For example, plaster can be moist-cured by using a suitable covering, such as polyethylene film, to provide a vapor barrier that will retain the moisture in the plaster. Moist curing can also be accomplished by using a fog spray of water. Tarpaulins or plywood barriers that deflect sunlight and wind will help to reduce evaporation rates.

An adequate temperature level is important to satisfactory curing. As the temperature drops, hydration slows and practically stops when the temperature approaches the freezing point. Therefore, portland cement plaster should not be applied to frozen surfaces, and frozen materials should not be used in the mix. Furthermore, in cold weather it may be necessary to heat the mixing water and the work area. ASTM C926 recommends a heated enclosure to maintain the air temperature above 40 deg. F. for 48 hours before plastering, while plastering, and for the duration of curing (at least 48 hours).

Ready Mixed Plaster

Ready mixed plaster is made with the same ingredients as conventional plaster, except it contains a set-controlling admixture that keeps the plaster workable for 2½ to 72 hours, depending on the admixture dosage and brand. The plaster is usually mixed at a central location, such as a ready mixed concrete plant, or in a mobile truck mixer. It is delivered to the jobsite in trowel-ready condition and stored at the jobsite in protected plastic containers (Fig. 7-11). The plaster is then delivered to the plasterer in smaller containers, such as 5-gallon pails, as needed (Fig. 7-12). The plaster is applied in the conventional manner. The primary advantages of ready mixed plaster are its uniformity (ingredients are accurately combined and mixed) and the relative ease of using the product. Also, jobsite mixing is not needed and plaster ingredients need not be stored at the jobsite. Ready mixed plaster is especially useful on large projects requiring large quantities of plaster and where mixing and storage space are limited. For more information, see the discussion on ready mixed mortar in Chapter 2.

Furred and Other Finishes

In addition to plaster, other finishes such as fiberboard, gypsum wallboard, and wood paneling are sometimes applied directly to concrete masonry wall surfaces

Fig. 7-11. Ready mixed plaster delivered to the project and stored in plastic containers.

Fig. 7-12. Ready mixed plaster is ready to use as needed. Here the plaster is transferred from the storage tub into 5-gallon pails for placement on the scaffolding near the plasterers.

(Chapter 3) and sometimes upon furring or insulation board as in exterior insulation and finish systems.

Furring, which may be wood or metal, ensures a definite air space ¾ to several inches wide between the masonry and the finish (Fig. 7-13). Furring may be necessary to:

1. Provide suitably plumb, true, and properly spaced supporting construction for a wall finish.
2. Eliminate capillary moisture transfer in exterior or below-grade concrete masonry walls, thus minimizing the likelihood of condensation on interior wall surfaces.
3. Improve thermal insulation.
4. Improve sound insulation.

The furring strips are fastened to the concrete masonry with cut nails, helically threaded concrete nails, or powder-actuated fasteners (Fig. 7-14). Adhesives can be used to attach wallboard directly to wood furring, although a few nails may be required until the adhesive has set.

The reader should realize that wood furring is combustible even if enclosed by masonry and a fire-resistant finish. Building codes sometimes prohibit such combustible construction.

Gypsum wallboard finishes consist of one or more layers of factory-fabricated gypsum board having a noncombustible gypsum core with surfaces and edges of paper or other materials. Gypsum plaster may be applied directly to concrete masonry or to lath that may or may not be furred. Gypsum products are not

Fig. 7-13. Nailing wallboard to wood furring strip.

Fig. 7-14. Wood furring strips are nailed to the mortar joints between the block.

Fig. 7-15. Typical PM type synthetic stucco system consists of *(from outside in):* a polymer-modified cement finish coat, a similar but thicker base coat with embedded fiberglass mesh, and XPS insulation board mechanically fastened to the substrate. Photo courtesy of Parex Inc.

recommended where significant exposure to moisture is expected. In such applications, portland cement plaster may be used or glass fiber mesh reinforced concrete backer board is available in many areas and is recommended for moisture exposures.

Typical materials used to increase thermal resistance of furred finishes include: (1) flexible fiber insulation (batts and blankets), (2) loose-fill insulation, (3) rigid plastic insulation, and (4) reflective-foil insulation attached to the back surface of wallboard. Mineral fiber blankets may be installed behind the lath or wallboard not only to improve thermal insulation but also to decrease sound transmission. Resilient clips for lath attachment may also be used to decrease sound transmission.

Exterior Insulation and Finish Systems

Exterior insulation and finish systems (EIFS) sold in the United States evolved from systems developed and used successfully in Europe for over 25 years. Used extensively on residential and commercial buildings, EIF systems can be used as an applied finish on concrete masonry walls (Fig. 7-15).

Also known as "synthetic stucco", EIF systems are exterior cladding assemblies consisting of a cementitious laminate that is wet-applied, usually in two coats, like plaster to rigid insulation board that is fastened to the wall with an adhesive, mechanical fasteners, or both. Lath, usually a fiberglass mesh, is embedded in the base coat. Expanded polystyrene (EPS) board is the insulation most often used, with extruded polystyrene (XPS) board a distant second; however, some manufacturers offer mineral wool or fiberglass insulation boards to create noncombustible EIF systems.

The composition and thickness of synthetic stucco in EIF systems varies widely from one manufacturer to another. In an attempt at standardization, the Exterior Insulation Manufacturers Association (EIMA) classified EIF systems as polymer based (PB); polymer modified, mineral based (PM); and mineral based (MB). The mineral base is portland cement, while the polymer usually is an acrylic. The PB systems are called thin coat, soft coat, or flexible finishes, while the PM systems are also known as thick coat, hard coat, or rigid finishes. The MB designation is for conventional or generic three-coat portland cement stucco from which all EIF systems have evolved.

The typical PB system consists of EPS insulation bonded to the substrate with adhesive; fiberglass mesh embedded in a base coat; and a finish coat. The base coat may be one part portland cement to one part polymer, applied about $1/16$ in. thick. The finish coat also is polymer-based and applied quite thin, often no thicker than the maximum aggregate particle in the coat's sand filler. PB systems are lightweight, flexible, and have low impact resistance. If the surface of concrete masonry is clean, a PB system can be applied with an adhesive, however a surface primer may be required. Surfaces with poor bond characteristics, such as painted walls, require mechanical fastening.

The typical PM system uses XPS insulation rather than EPS; is mechanically fastened rather than glued; and uses a base and finish coat that totals ¼ in. in thickness. PM systems are much tougher, rigid, and have high impact resistance. They require control joints similar to conventional stucco whereas the PB systems do not. However, all EIF systems require construction joints to accommodate stopping and starting places in a days work.

One of the main advantages claimed for EIF systems is that insulation is placed where it is most effective—on the building's exterior. PB systems tend to be more economical than PM systems because they use less expensive insulation and the stucco coats are thinner. However, PB systems offer less penetration resistance than PM systems. The finishing options with EIF systems are about the same as those for conventional portland cement plaster; that is, a wide range of colors and textures are available.

There are very few guidelines other than the manufacturers' for installation of EIF systems. ASTM is establishing committees to write standards and some guidelines are available from EIMA. With the rapid increase in use of synthetic stucco and lack of product standards, there have been reports of failures, especially cracking of the finish and sealant failures in soft-coat systems. There have been fewer problems with EIF systems that have used concrete block as the major supporting substrate (see Ref. 90). Concrete masonry walls make a superior substrate for EIF systems because they are very stable and not affected by moisture that may enter the wall. For more information on EIF systems, see References 90-92.

Chapter 8

VARIOUS APPLICATIONS OF CONCRETE MASONRY

The most common application of concrete masonry is for built-in-place walls for buildings of all kinds. However, there are a number of other common applications, as described in this chapter.

Fireplaces and Chimneys

Fireplaces and chimneys are important elements in the design and construction of homes. The fireplace can be a central feature for family social life, and the chimney often is a dominant and interesting architectural feature on the home exterior as well as a focal point of interior design. Accordingly, it is desirable that fireplaces and chimneys be esthetically pleasing as well as functional.

Various requirements for fireplaces and chimneys normally are set forth in local building codes, but they are usually for a single residential fireplace. In the event the chimney is multistory, extra wide, or extra high or there are multiple fireplaces and flues within the chimney, special design considerations are necessary. A fireplace can be located on one floor directly above one on the floor below, but each fireplace must have a separate flue. Fig. 8-2 shows a way to combine multilevel fireplaces into one chimney. Each flue takes off from the center of the smoke chamber.

The design and construction of an efficient, functional fireplace requires adherence to basic rules concerning fireplace location and the dimensions and placement of various component parts, keeping in mind the basic functions of a fireplace. These functions are: (1) to assure proper fuel combustion, (2) to deliver smoke and other products of combustion up the chimney, (3) to radiate the maximum amount of heat, and (4) to provide an attractive architectural feature. It must also afford simplicity and firesafety in construction.

Combustion and smoke delivery depend mainly upon the shape and dimensions of the combustion chamber, the proper location of the fireplace throat and the smoke shelf, and the ratio of the flue area to the area of the fireplace opening. The third objective, heat radiation, depends on the dimensions of the combustion chamber and proper slope and construction of angles and flue. Firesafety depends not only on the

Fig. 8-1. Attractive single-face fireplace of split-face concrete masonry units.

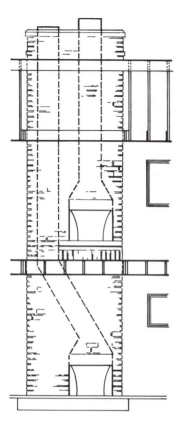

Fig. 8-2. Chimney with separate flues for fireplaces on two floors.

design of the fireplace and chimney but also on the ability of the masonry units to withstand high temperatures without warping, cracking, or deteriorating.

When properly designed, concrete masonry can also be used safely for chimneys for wood stove heating of residences (usually cabins). A dangerous error sometimes created by do-it-yourselfers is the omition of flue liners or stacks. Fireclay mortar instead of portland cement mortar should be used in firebrick masonry and between flue liners. Fireplace owners should also be aware that creosote, a byproduct of wood burning, can slowly attack concrete masonry if not removed periodically.

Types of Fireplaces

There are several types of fireplaces being used today, and the basic principles involved in their design and construction are the same. These types and some standard sizes found to work satisfactorily under most conditions are given in Fig. 8-3 and Table 8-1.

The *single-face* fireplace (pictured in Fig. 8-1) is the oldest and most common variety, and most standard design information is based on this type. The *multi-face* fireplace, used properly, is highly effective and

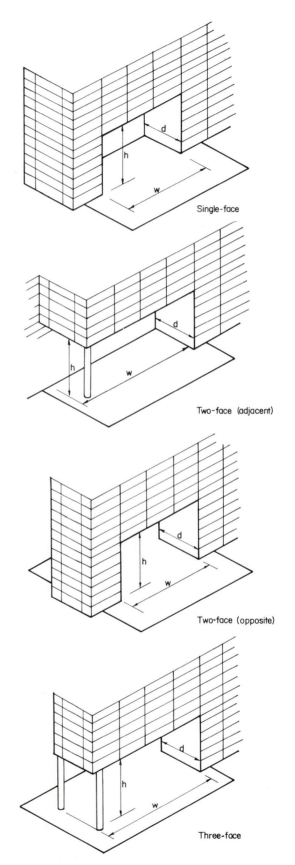

Fig. 8-3. Fireplace types with one to three faces.

Table 8-1. Fireplace Types and Standard Sizes*

Type**	Width (*w*), in.	Height (*h*), in.	Depth (*d*), in.	Area of fireplace opening, sq.in.	Nominal flue sizes (based on $^1/_{10}$ area of fireplace opening),† in.
Single-face	36	26	20	936	12×16
	40	28	22	1,120	12×16
	48	32	25	1,536	16×16
	60	32	25	1,920	16×20
Two-face (adjacent)	39	27	23	1,223	12×16
	46	27	23	1,388	16×16
	52	30	27	1,884	16×20
	64	30	27	2,085	16×20
Two-face (opposite)	32	21	30	1,344	16×16
	35	21	30	1,470	16×16
	42	21	30	1,764	16×20
	48	21	34	2,016	16×20
Three-face	39	21	30	1,638	16×16
	46	21	30	1,932	16×20
	52	21	34	2,184	20×20

*Adapted from Ref. 15.
**Fireplace types and dimensions *w, h,* and *d* are shown in Fig. 8-3.
†A requirement of the U.S. Federal Housing Administration if the chimney is 15 ft. high or over; $^1/_8$ ratio is used if chimney height is less than 15 ft. See Table 8-2 for nominal and actual flue sizes and inside areas of flue linings.

attractive, but it may present certain problems as to draft and opening size that usually must be solved on an individual basis. The *two-face (opposite* or *see through)* fireplace is not recommended by some governmental agencies. If used, this type of fireplace requires a fire screen of fire-resistant, tempered pyrex glass to be placed on one side to prevent fire from blowing out into the room.

Barbecues or outdoor fireplaces can discharge into a chimney attached to the house with a separate flue or, if desired, can be located separately from the house with its own chimney. Inexpensive, serviceable barbecues can be built of concrete masonry with minimum labor and time. The site selected should be sheltered from the wind, conveniently located between play and work areas, and afford adequate entertaining space. Details of a simple barbecue are shown in Fig. 8-4 and those of a more elaborate one in Fig. 8-5.

Fireplace Elements

Only the major elements of a fireplace are discussed below, but the details of a typical unreinforced concrete masonry fireplace are shown in Fig. 8-6. For more information see Reference 84.

Hearth

The floor of the fireplace is called the hearth. The inner part of the hearth is lined with firebrick and the outer hearth consists of noncombustible material such as firebrick, clay brick, natural stone, concrete brick, concrete block, or reinforced concrete. The outer hearth is supported on concrete that may be part of a ground floor or a cantilevered section of the slab supporting the inner hearth.

Lintel

The fireplace lintel is the horizontal member that supports the front face or mantel of the fireplace above the opening. It may be made of reinforced masonry or a steel angle, the same as other lintels discussed in Chapter 4. However, it is possible to eliminate the lintel by use of a masonry arch.

Firebox

The combustion chamber where the fire occurs is called the firebox. Its sidewalls are slanted slightly to radiate heat into the room, and its rear wall is curved or inclined to provide an upward draft to the throat (described below) and to help radiate heat into the room.

Unless the firebox is of the metal preformed type (at least ¼ in. thick), it should be constructed with firebrick that is at least 2 in. thick. The firebrick should be laid with thin joints of fireclay (refractory) mortar, not portland or masonry cement mortars, because they disintegrate under direct fire exposure. The back and sidewalls of the firebox should be at least 8 in. thick to

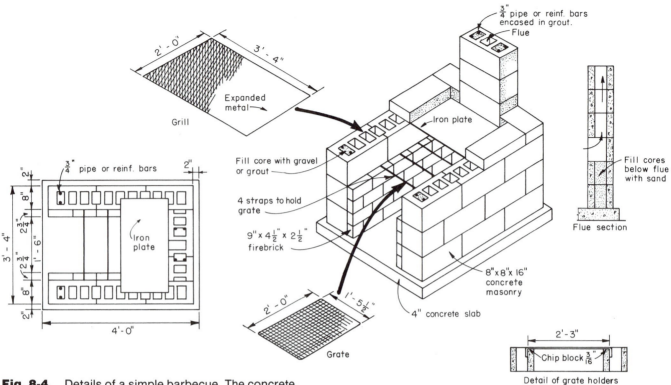

Fig. 8-4. Details of a simple barbecue. The concrete masonry flue must be kept at least 24 in. clear of any combustible construction.

support the weight of the chimney above. Woodwork and other combustible material must not be placed within 6 in. of the fireplace opening.

In the generally accepted method of construction, the fireplace is laid out and its back constructed to a scaffold height of approximately 5 ft. before the firebox is constructed and backfilled with mortar and brick scraps. Masons should not backfill solidly behind the firebox wall but slush the mortar loosely to allow for some expansion of the firebox.

Throat

The throat of a fireplace is the slot-like opening directly above the firebox through which flames, smoke, and other combustion products pass into the smoke chamber. Because of its effect on the draft, the throat must be carefully designed to be not less than 6 in. and preferably 8 in. above the highest point of the fireplace opening, as shown in Fig. 8-6.

The sloping or inclined back of the firebox should extend to the same height as the throat and form the support for the hinge of a metal damper placed in the throat. The damper extends the full width of the fireplace opening and preferably opens both upward and backward.

Smoke Chamber

The smoke chamber acts as a funnel to compress the smoke and gases from the fire so that they will enter the chimney flue above. The most convenient shape of a smoke chamber would be symmetrical with respect to the center line of the firebox. However, when two- or three-faced fireplaces are constructed, the smoke shelf may be located in the area adjacent to but not over the firebox. The back of the smoke chamber is usually vertical and its other walls are inclined upward to meet the bottom of the chimney flue lining. If the wall thickness is less than 8 in. of solid masonry, the smoke chamber should be parged with ¾ in. of fireclay brought to a smooth texture. Metal lining plates are available to give the chamber its proper form, provide smooth surfaces, and simplify masonry construction.

Chimney Elements

A fireplace chimney serves a dual purpose: to create a draft and to dispose of the products of combustion. Careful consideration must be given to chimney design and erection in order to assure efficient operation and freedom from fire hazards. Some of the requirements for chimney construction are mentioned below. For more information, see Reference 84.

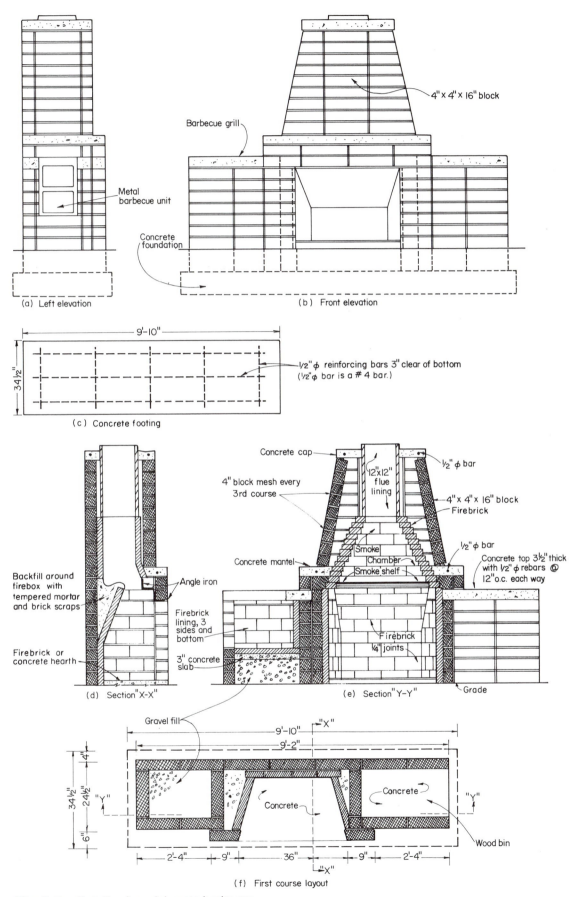

(a) Left elevation

Metal barbecue unit

Barbecue grill

Concrete foundation

4" x 4" x 16" block

(b) Front elevation

9'-10"

34½"

½" φ reinforcing bars 3" clear of bottom
(½" φ bar is a #4 bar.)

(c) Concrete footing

Concrete cap

4" block mesh every 3rd course

Concrete mantel

Backfill around firebox with tempered mortar and brick scraps

Firebrick or concrete hearth

Firebrick lining, 3 sides and bottom

3" concrete slab

Angle iron

(d) Section "X-X"

12"x12" flue lining

½" φ bar

4" x 4" x 16" block
Firebrick

½" φ bar

Concrete top 3½" thick with ½" φ rebars @ 12" o.c. each way

Smoke Chamber

Smoke shelf

Firebrick
¼" joints

Grade

(e) Section "Y-Y"

Gravel fill

9'-10"

9'-2"

4"

34½" 24½"

6"

Concrete

Concrete

Wood bin

2'-4"   9"   36"   9"   2'-4"

"X"

"Y"

"Y"

"X"

(f) First course layout

Fig. 8-5. Details of an elaborate barbecue.

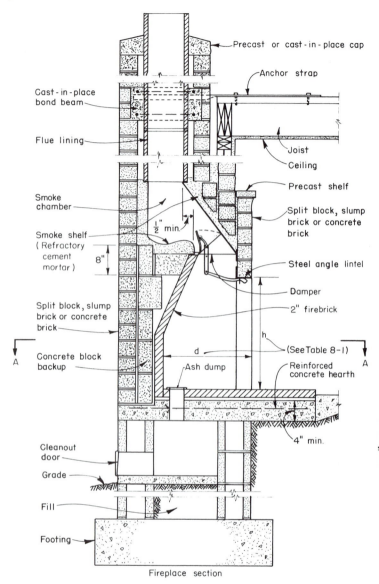

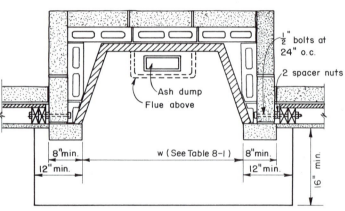

NOTE: The drawings and text for fireplace and chimney elements do not constitute complete working details, specifications, or instructions for construction. In the interest of health and firesafety, local and regional codes should be consulted.

The following document is recommended for study when connecting or installing wood stoves, fireplace inserts, fireplace stoves, heaters, and furnaces:
- NFPA 211, Standard for Chimneys, Fireplaces, Vents, and Solid Fuel Burning Appliances, National Fire Protection Association, see Ref. 84.

Fig. 8-6. Unreinforced concrete masonry fireplace and chimney. Some building codes require that the concrete masonry units be solid units (see definition in "Hollow and Solid Units," Chapter 1).

To prevent upward draft from being neutralized by downward air currents, the chimney should be extended at least 3 ft. above a flat roof, 2 ft. above the ridge of a pitched roof, or 2 ft. above any part of the roof within a 10-ft. radius of the chimney. If the chimney does not draw well, increasing its height will improve the draft. A typical unreinforced concrete masonry chimney is shown with the fireplace in Fig. 8-6.

Foundation

Usually made of concrete, the foundation for a chimney is designed to support the weight of the chimney and any additional load. Because of the large mass and weight imposed, it is important that the unit bearing pressure beneath the chimney foundation be approximately equal to that beneath the house foundation; this will minimize the possibility of differential settlement. The chimney foundation is generally unreinforced, with only the chimney reinforcement (where required by local building codes) extending from it.

The footing thickness should be at least 8 in. for 16-in. square chimney units, a typical concrete footing is 30 in. square and 12 in. thick. The bottom of the footing should be at least 18 in. below grade and extend below the frost line.

Chimney Flue

A fireplace chimney flue must have the correct area and shape to produce a proper draft.* Relatively high velocities of smoke through the throat and flue are desirable. Velocity is affected by the flue area, the firebox opening area, and the chimney height.

Generally the required cross-sectional area of the flue should be approximately ¹⁄₁₀ of the area of the fireplace opening. However, since some codes may specify ⅛ or ½ under varying conditions, the local building department should be consulted. Typical sizes of fireplace flues and flue linings are given in Tables 8-1 and 8-2.

A fireplace chimney can contain more than one flue, but each flue must be built as a separate unit entirely free from the other flues or openings. Flue walls should have all joints completely filled with mortar. All chimney flues should be lined. Clay flue liners are the common requirement and are covered by ASTM C315, Standard Specification for Clay Flue Linings. Firebrick can also be used. Concrete flue liners made with perlite aggregate and portland cement have been approved in Research Committee Recommendation Report No. 2602 of the International Conference of Building Officials (ICBO).

Table 8-2. Clay Flue Lining Sizes*

Nominal size, in.	Manufactured size (modular), in.**	Inside area, sq.in.
4 × 8	3½ × 7½	15
4 × 12	3½ × 11½	20
4 × 16	3½ × 15½	27
8 × 8	7½ × 7½	35
8 × 12	7½ × 11½	57
8 × 16	7½ × 15½	74
12 × 12	11½ × 11½	87
12 × 16	11½ × 15½	120
16 × 16	15½ × 15½	162
16 × 20	15½ × 19½	208
20 × 20	19½ × 19½	262
20 × 24	19½ × 23½	320
24 × 24	23½ × 23½	385

*Source: Clay Flue Lining Institute
**Actual dimensions may vary somewhat, but the flue lining must fit into a rectangle corresponding to the nominal flue size.

---

*This discussion of flues deals only with residential fireplace chimneys. Commercial and industrial chimneys have more stringent requirements.

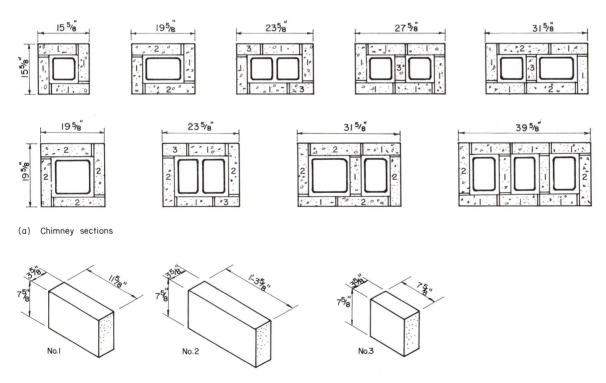

(a) Chimney sections

(b) Solid units for use with clay flue lining

Fig. 8-7. Residential concrete masonry chimney sections. See Table 8-2 for dimensions of clay flue linings.

Flue linings should start at the top of the smoke chamber and extend continuously to 4 to 8 in. above the chimney cap. The chimney walls are constructed around the flue lining segments, which are embedded one upon the other in a refractory mortar such as fireclay, and left smooth on the inside of the lining. Liners should be separated from the chimney wall by at least ½ in. of air space, and the space between the liner and masonry is not filled; only enough mortar should be used to make a good joint and hold the liners in position.The air space allows the liner to expand and contract independently of the wall without affecting the chimney. If the void is inadvertently filled with mortar, extensive cracking can occur throughout the chimney.

The minimum air space between the chimney and any framing material should be 2 in. Firestopping in between must be made of noncombustible material. Firestopping can be galvanized steel (26 guage minimum) or sheet materials not more than ½ in. thick.

Modular-size solid units (Fig. 8-7) can be combined with modular-size flue lining for modular-size concrete masonry construction. Minimum wall thickness measured from the outside of the flue lining should be 4 in. nominal. The exposed joints inside the flue are struck smooth and the interior surface is not plastered.

Smoke pipe connections should enter the side of the flue at a thimble or flue ring that is built of fireclay or firebrick set with fireclay mortar. The metal smoke pipe should not extend beyond the inside face of the flue, and the top of the smoke pipe should be not less than 18 in. below the ceiling. No wood or combustible materials should be placed within 6 in. of the thimble.

When a chimney contains more than two flues, they should be separated into groups of one or two flues by 4-in.-thick masonry bonded into the chimney wall, or the joints of the adjacent flue linings should be staggered at least 7 in. The tops of the flues should have a height difference of 2 to 12 in. to prevent smoke from pouring from one flue into another. A fireplace on an upper level should have the top of its flue higher than the flue of a fireplace on a lower level as shown in Fig. 8-2.

For reasons of appearance, chimneys are often built to the same widths as attached fireplaces, and these wide chimneys sometimes contain only a single flue. It can be located anywhere within the chimney. Consideration should be given to reinforcing a wide chimney wall against lateral forces (see discussion "Reinforcement and Chimney Anchorage").

Practically any size or shape of single- or multiple-flue chimney can be constructed with only three different sizes of solid concrete block units (designated Nos. 1, 2, and 3 in Fig. 8-7).

Chimneys should be built as nearly vertical as possible, but a slope is allowed if the full area of the flue is maintained throughout its length. When a slope from the vertical is required in the flue for design reasons, it

should not exceed 7 in. to the foot or 30 deg. Where offsets or bends are necessary, they should be formed by mitering both ends of abutting flue liner sections equally; this prevents reduction of the flue area.

Chimney Cap and Hood

The top of the chimney wall should be protected by a concrete cap conforming with the architectural design of the building. The cap should slope not only to prevent water from running down next to the flue lining and into the fireplace but also to prevent standing water from creating frost or moisture problems. In addition, since chimney flues should project 4 to 8 in. above the cap, a sloping cap improves draft from the flue as well as the smoke exhaust characteristics of the chimney. If the cap projects beyond the chimney wall a few inches, a drip slot in its lower edge should be included to help keep the wall dry and clean. The flue liner should also be structurally independent of the chimney cap. A sheet metal bond breaker is helpful for separating the cap and liner.

A chimney hood gives a finished touch to the silhouette of the building. It protects the chimney and fireplace from rain and snow and, when the building is located below adjoining buildings, trees, and other obstacles, prevents downdrafts. It must have at least two sides open, with the open areas larger than the flue area. A simple concrete hood is shown in Fig. 8-8.

Concrete chimney hoods should be reinforced with steel bars or welded-wire fabric. If a hood projects from a chimney wall, a drip slot under the edge is included. Also, the openings are sometimes enclosed with heavy screening to keep out small animals and birds, but insurance company regulations on screening should be checked.

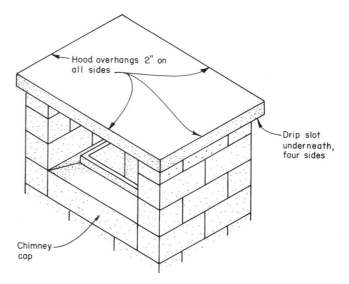

Fig. 8-8. A simple chimney hood keeps rain and snow out, prevents downdrafts, and improves appearance.

Reinforcement and Chimney Anchorage

Depending on local building codes, fireplaces and chimneys have to be reinforced in areas subject to earthquakes or high wind loads. A typical reinforced concrete masonry fireplace and chimney are shown in Fig. 8-9.

The reinforcement, consisting of at least four ½-in.-diameter deformed vertical bars, should extend the full height of the chimney and be tied into the footing and chimney cap. Also, the bars should be tied horizontally with ¼-in.-diameter ties at not more than 18-in. intervals. If the width of a reinforced chimney exceeds 40 in., two additional ½-in. vertical bars should be provided for each additional flue incorporated into the chimney, or for each additional 40-in. width or fraction thereof.

All chimneys not located entirely within the exterior walls of a residence should be anchored to the building at each floor or ceiling line 6 ft. or more above grade and at the roof line. The anchors should consist of ¼-in. steel straps wrapped around vertical reinforcement or chimney flues, as shown in Fig. 8-6. Each end of the strap is attached to the structural framework of the building with six 16d nails, two ½-in.-diameter bolts, or two ⅜-in.-diameter by 3-in.-long lag screws. Reinforced chimneys must have equivalent anchorage, as shown in Fig. 8-9.

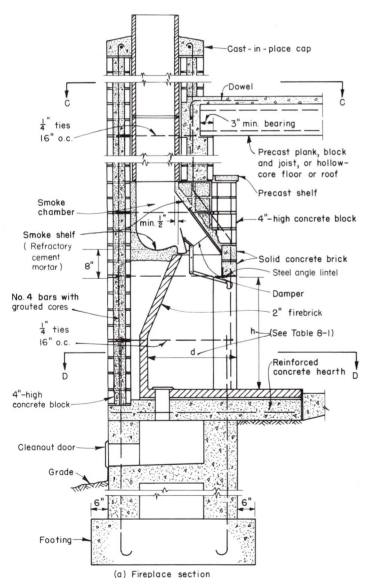

(a) Fireplace section

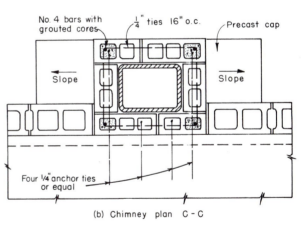

(b) Chimney plan C - C

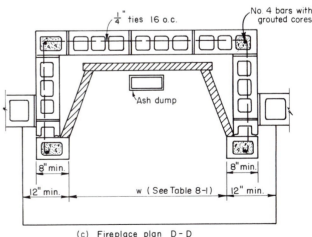

(c) Fireplace plan D - D

Fig. 8-9. Reinforced concrete masonry fireplace and chimney.

When a chimney extends considerably above the roof level, an intermediate lateral support or tie is often placed between the roof line and the chimney.

Screen Walls

Concrete masonry screen walls (walls constructed with over 25% exposed open areas) are functional, decorative elements. They combine privacy with a view, interior light with shade and solar heat reduction, and airy comfort with wind control. Curtain walls, sun screens, decorative veneers, room dividers, and fences are just a few of the many applications of the concrete masonry screen wall.

Materials

With conventional concrete block or the specially designed screen wall units or grille block, concrete masonry offers a new dimension in screen wall design. The many sizes, shapes, colors, patterns, and textures available help in creating imaginative designs. Several units may make up a design, or each screen wall unit may constitute a design in itself.

Although the number of designs for concrete masonry screen units is virtually unlimited, it is advisable to check on availability of any specific unit during the early planning stage. Some designs are available only in certain localities and others are restricted by patent or copyright. A few screen units are shown in Fig. 1-24.

Screen units should be of high quality, even though they are not often used in load-bearing construction. When tested with their hollow cells parallel to the direction of a test load, screen units should have a minimum compressive strength of 1,000 psi on the gross area.

Type S mortar should be used for exterior screen walls and where the screen walls are required to carry any vertical load. For interior non-load-bearing walls, the mortar may be Type O, or N. Grout for embedding steel reinforcement in horizontal or vertical cells should comply with the criteria described in Chapter 2.

Design and Construction

Care must be taken that screen walls are stable and safe, although from a design standpoint they are seldom required to support more load than their own weight. They can be designed as load-bearing walls, but such construction is not permitted by some building codes. In any case, extra attention to design of screen walls for wind forces is warranted because of and despite the relatively high percentage of open areas. The wall openings are created by using screen

Fig. 8-10. A concrete grille block wall makes a pleasing backdrop and screen.

units with decorative openings or, occasionally, by using conventional concrete block with intermittent vertical mortar joints, that is, by leaving openings in the wall in lieu of vertical mortar joints.

Screen walls should be designed to resist all the horizontal forces that can be expected. Stability is provided by:

1. Using a framing system capable of carrying horizontal forces into the ground.
2. Employing adequate connection or anchorage of screen walls to the framing system.
3. Limiting the clear span.
4. Incorporating vertical reinforcement and horizontal joint reinforcement.

Partitions built with the screen units are usually designed as non-load-bearing panels, with primary consideration given to adequate anchorage at panel ends and/or top edge, depending on where lateral support is furnished. Lateral design pressure should be at least 5 psf. Lateral support for screen walls may be obtained from cross walls, piers, columns, posts, or buttresses for the horizontal spans—and from floors, shelf angles, roofs, bond beams, or spandrel beams for screen walls spanning the vertical direction. The structural frame for a screen wall may consist of reinforced concrete masonry columns, pilasters, and beams, or may incorporate structural steel members as shown in Fig. 8-11. Screen wall framing methods may also be similar to those used for fences (see next section).

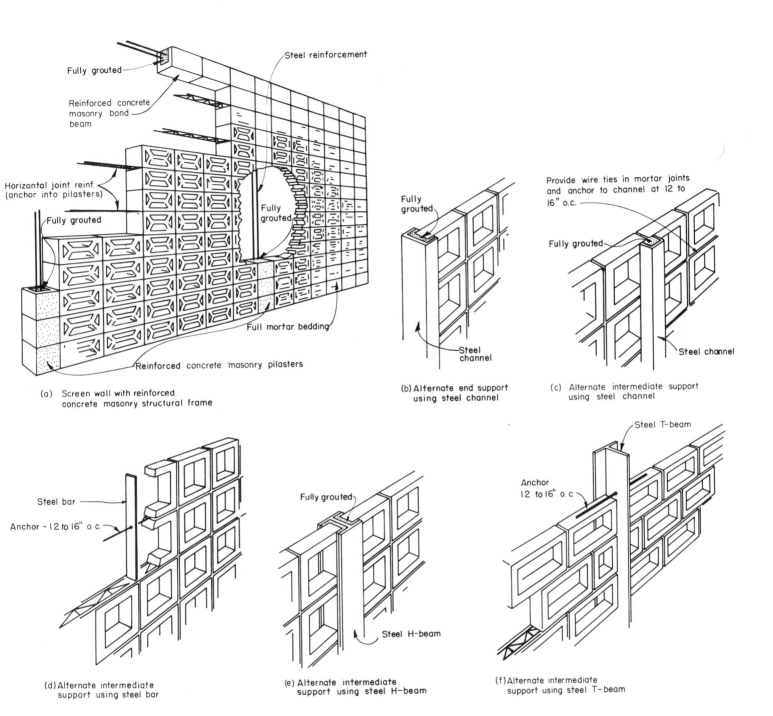

Fig. 8-11. Framing methods for screen walls. (Adapted from Ref. 9.)

When designed as veneer, the concrete masonry screen wall is attached to a structural backing with wire ties or sheet metal anchors in the same manner as used for other types of masonry veneer.

A non-load-bearing screen block panel may be used to fill an opening in a load-bearing masonry wall. In this case the panel is restrained on all four sides. Joint reinforcement is placed in the horizontal joints to anchor the panel into the wall. For an exterior wall, a panel is limited to 144 sq.ft. of wall surface or 15 ft. in any direction; for an interior wall, 250 sq.ft. or 25 ft. in any direction. The lintel, sill, and jamb of the panel opening should be designed the same as for a window opening.

Non-load-bearing screen walls should have a minimum nominal thickness of 4 in. and a maximum clear span of 36 times the nominal thickness. For load-bearing screen walls, the minimum thickness should be increased to 6 in. The maximum span can be measured vertically or horizontally, but need not be limited in both directions.

Screen walls are usually capable of carrying their own weight up to 20 ft. in height, but above that height they must be supported vertically not more than every 12 ft. When screen walls support vertical loads, the allowable compressive stress should be limited to 50 psi on the gross area. In some instances the compressive stress at the base of a non-load-bearing screen wall will govern maximum unsupported height. Where screen block are not laid in a continuous mortar bed (intermittent bond), the allowable stresses should be reduced in proportion to the reduction in the mortar-bedded area.

Due to the somewhat fragile nature of screen walls, the use of steel reinforcement is recommended wherever it can be embedded in mortar joints and bond-beam courses, or grouted into continuous vertical or horizontal cavities. When reinforced joints are used, the thickness of the mortar joint should be a minimum of twice the diameter of the reinforcement.

For exterior screen walls, joints and connections should be constructed as fully watertight as possible. The mortar joints should be constructed according to

Fig. 8-13. Screen block used to create a dignified property-line fence.

the best construction practices. In addition, when hollow masonry units are laid with their cores vertical, the top course should be capped to prevent the entrance of water into the wall interior.

Garden Walls and Fences

Garden walls and fences of concrete masonry can take on many delightful forms, enhancing the landscape. They are built with solid or screen block and with concrete brick or half block. If a garden wall has more than 25% open areas, it may be considered a fence. Fence framing methods are shown in Fig. 8-14.

Fences and garden walls should be able to safely withstand wind loads of at least 5 psf, and most city codes specify resistance to 20-psi pressure. Pressures and corresponding wind gust velocities are:

Pressure, psf	Wind gust velocity, mph
5	40
10	57
15	69
20	80

In hurricane areas higher wind-pressure resistance is needed.

Examples of reinforced garden walls or fences are shown in Fig. 8-15. Without reinforcement, high and

Fig. 8-12. Garden wall with regular and screen units.

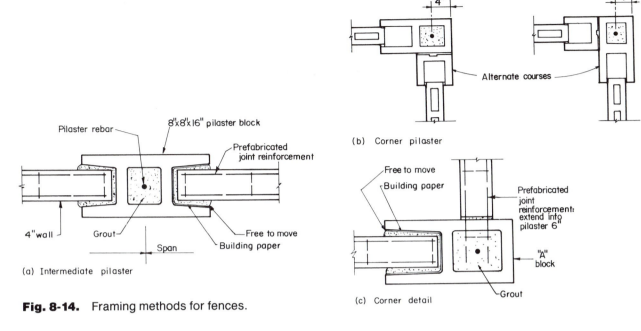

(b) Corner pilaster

(c) Corner detail

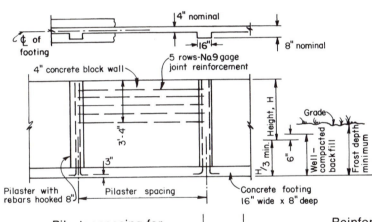

(a) Intermediate pilaster

Fig. 8-14. Framing methods for fences.

Pilaster spacing for wind pressure					Reinforcement for wind pressure			
5 psf	10 psf	15 psf	20 psf	H	5 psf	10 psf	15 psf	20 psf
19'4"	14'0"	11'4"	10'0"	4'0"	1—No. 3	1—No. 4	1—No. 5	2—No. 4
18'0"	12'8"	10'8"	9'4"	5'0"	1—No. 3	1—No. 5	2—No. 4	2—No. 5
15'4"	10'8"	8'8"	8'0"	6'0"	1—No. 4	1—No. 5	2—No. 5	2—No. 5

(a) Wall or fence with pilasters

	Reinforcement for wind pressure			
H	5 psf	10 psf	15 psf	20 psf
4'0"	1—No. 3	1—No. 3	1—No. 4	1—No. 4
5'0"	1—No. 3	1—No. 4	1—No. 5	1—No. 5
6'0"	1—No. 3	1—No. 4	1—No. 5	2—No. 4

(b) Wall or fence without pilasters

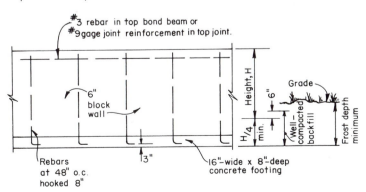

Fig. 8-15. Reinforced garden walls or fences.

Fig. 8-16. A tall garden fence can be simply constructed on the serpentine principle using conventional two-core block.

straight garden walls or fences lack vertical tensile strength and are unstable in strong winds. For a 57-mph peak wind velocity, the safe height of a straight 6-in. block wall is only 3 ft. 6 in.; for a straight 8-in. wall, only 5 ft. 6 in. (see Fig. 6-69 in Chapter 6).

The serpentine wall or fence (Fig. 8-16) is a welcome change from the straight lines frequently seen on our modern landscapes. Undulating curves and "folded plates" give this type of wall superb stability from the foundation up with no need for reinforcement.

Fig. 8-17 shows sample designs of serpentine walls based on proportions found safe for wind gusts with pressures up to 20 psf; the horizontal radius should not exceed twice the height, which in turn should not be more than twice the width. A limiting height of 15 times the thickness is recommended. The free end(s) of the serpentine wall should have additional support, such as a pilaster or a short-radius return as shown in Fig. 8-17a.

Concrete masonry wall foundations may not be durable if they frequently become frozen while saturated, as noted in the discussion of "Durability" in Chapter 2. In cold climates, therefore, the wall foundations should be constructed with cast-in-place concrete.

Retaining Walls

Concrete masonry retaining walls can have visual beauty along with the required structural strength to resist imposed vertical and lateral loads (Fig. 8-18). Because the purpose of a retaining wall is to hold back a mass of soil or other material, the design of the wall is affected by the earth's configuration—whether the earth surface behind the wall is horizontal or inclined. Design is also affected by any additional loading (surcharge), such as from a vehicle or equipment passing near the top of the wall, which causes horizontal thrust on the retaining wall.

Types

There are three basic types of concrete masonry retaining walls to consider: gravity, cantilever, and counterfort or buttressed walls.

Gravity Walls

A gravity wall depends upon its own mass for stability. Basically, it is massive masonry laid so that little or no tension stress occurs in the wall under loading, and its cross-sectional shape is usually trapezoidal, as shown in Fig. 8-19. Since a retaining wall of the gravity type ordinarily has a base thickness equal to one-half to three-fourths the wall height, it is usually more economical to build a large retaining wall (more than a few feet in height) as a cantilever wall.

Many types of decorative units are available for constructing simple gravity retaining walls. Some units are interlocking and some also are filled with soil or granular fill. Most of these units do not use mortared joints (Fig. 8-20).

Cantilever Walls

A cantilever wall usually has a cross-sectional shape of an inverted T (Fig. 8-21), with the stem located towards the rear of the footing if soil bearing stresses are critical and towards the front or toe if sliding is critical. An L-shaped cantilever wall is used along a property line or in other situations where it is impossible to provide a toe; for such a wall, bearing pressure is usually high.

With either shape of cantilever wall, the reinforced masonry wall portion or stem performs structurally as a cantilever from the cast-in-place concrete footing. The portion of backfill directly above the footing contributes to the mass required for stability, and the concrete masonry is reinforced along the back face where the loading induces tensile stress. The functions of the footing are to hold the stem in position and to resist the forces of the stem (the sliding, overturning, and vertical pressures created by loading), transferring them to the soil.

Counterfort or Buttressed Walls

These retaining walls are similar to cantilever walls except that they span horizontally between vertical

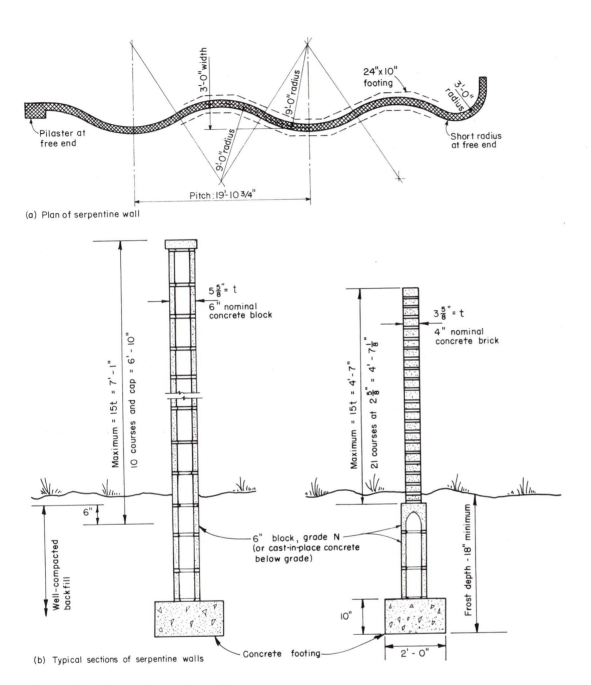

(a) Plan of serpentine wall

(b) Typical sections of serpentine walls

Fig. 8-17. Serpentine garden walls.

Fig. 8-18. Concrete masonry retaining wall with split-faced units.

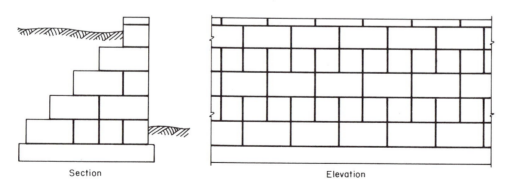

Section

Elevation

Fig. 8-19. Gravity retaining wall.

Fig. 8-20. Decorative interlocking concrete masonry retaining wall.

Table 8-3. Safe Bearing Pressures of Soils

Material	Bearing capacity, psf
Clay	2,000
Sand and clay mixed	4,000
Alluvium and silt	5,000
Hard clay and firm, compressed sand	8,000
Fine sand	9,000
Compacted and cemented sand	10,000

supports. Supports at the back of the wall are known as *counterforts* (Fig. 8-22), and those supports exposed at the front are called *buttresses* (Fig. 8-23).

A small degree of forward or outward tilt under service conditions is difficult to avoid with any type of retaining wall. It is therefore good practice to batter (slope) the front of the wall slightly to offset this tilt and avoid the illusion of instability. A batter on the order of ½ in. per foot is commonly used.

The selection of a particular type of retaining wall for cost and efficiency depends on the wall size, loads, soil conditions, and site location. The cantilever type of wall has a slightly lower toe pressure than the gravity type and thus may be desirable where soil bearing capacity is low. However, the gravity wall has greater resistance to sliding because of its greater weight.

It is good practice, both for design and construction of retaining walls, to use the services of an engineer who has experience with the local soil involved. He can design the cross-sectional dimensions of the wall and footing so that the computed pressure does not exceed the safe bearing value of the soil.* Table 8-3 gives safe values for different soils.

Construction

Footing

The footing for a retaining wall should be placed on firm, undisturbed soil. In areas where freezing temper-

atures are expected, the base of the footing is placed below the frost line. Where soil under the footing consists of soft or silty clay, 4 to 6 in. of consolidated granular fill can be placed under the footing slab to assure firm support and to increase the frictional resistance between the footing and the ground. This friction determines resistance to horizontal sliding of the wall.

Often a lug or key under the footing is provided to assist in resisting the tendency to slide (Fig. 8-24). The same effect is achieved by requiring that the footing be well below the excavated surface in undisturbed soil, particularly if the wall is higher than 7 ft. above the footing.

Dowels to connect the wall to the footing are located so that they will be adjacent to the vertical wall reinforcement when it is placed. With a small longitudinal bar provided along the dowel line near the top of the footing, the dowels can be accurately spaced and securely tied in the correct position.

The top of the concrete footing in the area under the masonry is roughened while the concrete is still fresh. Otherwise, a 1-in.-deep by 4-in.-wide keyway is provided to improve shear bond at the joint between the wall and the footing.

Grouting and Reinforcement

The first course of block on the footing is laid in a full mortar bed. The remaining courses may then be laid with mortar coverage on the face shells and on any web between a core to be grouted and a core not to be grouted. However, there appears to be little advantage in grouting only those cores containing reinforcement. If all cores are grouted, the small additional grouting material and labor costs are offset to some extent by eliminating the necessity of buttering cross webs that adjoin the cores to be grouted.

The materials and procedures previously recommended for reinforced, grouted masonry walls (Chapter 6) should be followed in the construction of retaining walls. It is necessary to provide some horizontal steel reinforcement to distribute stresses that occur when the wall expands or contracts. The amount of horizontal reinforcement needed is, to a large extent,

_____
*See Refs. 10 and 17 for retaining wall design examples.

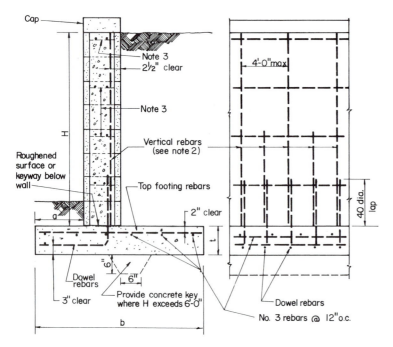

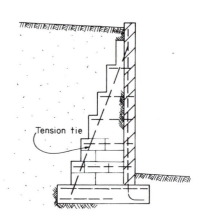

Fig. 8-22. Counterfort retaining wall.

Block width	H	a	b	t	Dowel and vertical reinforcement	Top footing reinforcement
8″	3′4″	12″	2′8″	9″	No. 3 @32″ oc	No. 3 @27″ oc
	4′0″	12″	3′0″	9″	No. 4 @32″ oc	No. 3 @27″ oc
	4′8″	12″	3′3″	10″	No. 5 @32″ oc	No. 3 @27″ oc
	5′4″	14″	3′8″	10″	No. 4 @16″ oc	No. 4 @30″ oc
	6′0″	15″	4′2″	12″	No. 6 @24″ oc	No. 4 @25″ oc
12″	3′4″	12″	2′8″	9″	No. 3 @32″ oc	No. 3 @27″ oc
	4′0″	12″	3′0″	9″	No. 3 @32″ oc	No. 3 @27″ oc
	4′8″	12″	3′3″	10″	No. 4 @32″ oc	No. 3 @27″ oc
	5′4″	14″	3′8″	10″	No. 4 @24″ oc	No. 3 @25″ oc
	6′0″	15″	4′2″	12″	No. 4 @16″ oc	No. 4 @30″ oc
	6′8″	16″	4′6″	12″	No. 6 @24″ oc	No. 4 @22″ oc
	7′4″	18″	4′10″	12″	No. 7 @32″ oc	No. 5 @26″ oc
	8′0″	20″	5′4″	12″	No. 7 @24″ oc	No. 5 @21″ oc
	8′8″	22″	5′10″	14″	No. 7 @16″ oc	No. 6 @26″ oc
	9′4″	24″	6′4″	14″	No. 8 @ 8″ oc	No. 6 @21″ oc

General notes:
1. Reinforcing bars should have standard deformations and a yield strength of 40,000 psi.
2. Alternate vertical reinforcing bars may be terminated at the midheight of the wall. Every third bar may be terminated at the upper third-point of the wall height.
3. The wall should have horizontal joint reinforcement at every course or else a horizontal bond beam with two No. 4 bars every 16 in.
4. Weight of assumed soil backfill (granular soil with conspicuous clay content) is 100 pcf and equivalent fluid pressure is 45 pcf. There is no surcharge and maximum soil bearing pressure is 1,500 psf.

Fig. 8-21. Cantilever retaining wall. (Adapted from Ref. 10.)

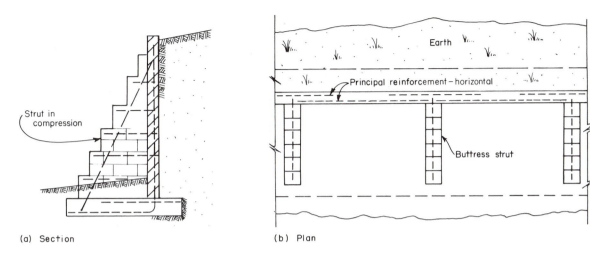

(a) Section

(b) Plan

Fig. 8-23. Buttressed retaining wall.

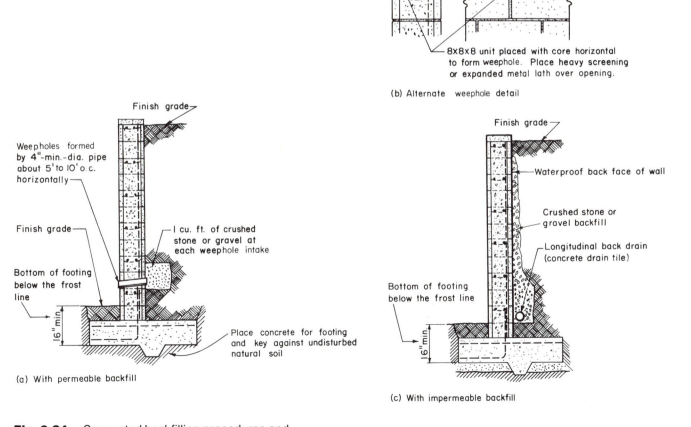

Crushed stone or gravel

8x8x8 unit

8x8x8 unit placed with core horizontal to form weephole. Place heavy screening or expanded metal lath over opening.

(b) Alternate weephole detail

Finish grade

Weepholes formed by 4"-min.-dia. pipe about 5' to 10' o.c. horizontally

Finish grade

Bottom of footing below the frost line

6" min.

I cu. ft. of crushed stone or gravel at each weephole intake

Place concrete for footing and key against undisturbed natural soil

(a) With permeable backfill

Finish grade

Waterproof back face of wall

Crushed stone or gravel backfill

Longitudinal back drain (concrete drain tile)

Bottom of footing below the frost line

6" min.

(c) With impermeable backfill

Fig. 8-24. Suggested backfilling procedures and drainage provisions for retaining walls.

dependent on climatic conditions. For moderate conditions and 8-in. walls, bond beams with two No. 4 bars should be placed in the top course and in intermediate courses at 16 in. on centers. For 12-in. walls, the top bond beam should contain two No. 5 bars and the intermediate bond beams should have two No. 4 bars. If desired, horizontal joint reinforcmeent may be placed in each joint (8 in. on center) and the bond beams omitted.

Drainage

Provisions must be made to prevent accumulation of water behind a retaining wall. Water accumulation causes increased soil pressure, seepage and, in areas subject to frost action, expansive forces of considerable magnitude near the top of the wall.

As shown in Fig. 8-24a, 4-in-diameter weepholes spaced 5 to 10 ft. along the base of the wall should provide sufficient drainage of permeable backfill soils. An alternate weephole detail appears in Fig. 8-24b. In another alternate method, the mortar is left out of the head joints in the first or second course and about 1 cu.ft. of gravel or crushed stone is placed around the intake for each weephole. Ideally, drains should be placed in all ungrouted cells in the first course.

Where unusual conditions such as heavy, prolonged rains will be encountered, seepage through weepholes may cause the ground in front of the wall and under the toe of the footing to become saturated and lose some of its bearing capacity. This undesirable condition can be avoided by installing a continuous longitudinal drain (Fig. 8-24c) that is surrounded by crushed stone or gravel near the base. Extending along the full length of the back of the wall, this drain should have outlets located beyond the ends of the wall, thus eliminating any need for weepholes. With impermeable soil and conditions tending to create excessive amounts of

water in the backfill, or in areas of frequent freezing and thawing, it is advisable to provide a continuous back drain (a vertical layer of crushed stone or gravel covering the entire back of the wall) in addition to a longitudinal drain (see Fig. 8-24c).

Other Provisions

The top of a concrete masonry retaining wall should be capped or otherwise protected to prevent entry of water into unfilled hollow cores and spaces. Climate and type of construction will determine the need for waterproofing the back face of the wall. Since saturated mortar may not be durable in areas subject to frequent freezing and thawing, waterproofing is recommended when backfill material is relatively impermeable; it is also recommended to reduce unsightly efflorescence or leaching on the wall.

A long retaining wall should be broken into panels by means of vertical control joints, as discussed in Chapter 4. The joints should resist shear and other lateral forces in order to maintain alignment of adjacent wall sections while permitting longitudinal movement (Fig. 8-25). In some cases, to prevent seepage through a joint, it may be advisable to cover the joint with a strip of waterproofing membrane on the back of the wall.

Care should be taken when backfilling against a retaining wall. Backfilling should not be permitted until at least 7 days after grouting. It is good practice to build up the backfill material all along the wall at a rate as nearly uniform as practicable. If heavy equipment is used in backfilling a wall designed to resist only earth pressure, such equipment should not approach the back of the wall closer than a distance equal to the height of the wall. Care should also be taken to avoid large impacting forces on the wall such as those caused by a large mass of moving earth or large stones

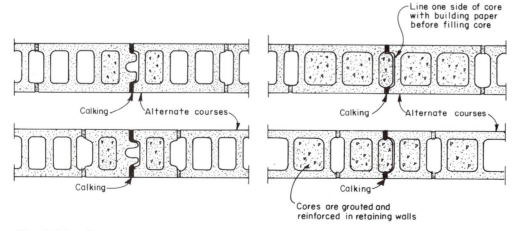

Fig. 8-25. Shear-resisting control joints for retaining walls.

Where the finished grade at the back of a retaining wall is level or nearly so, a fence or railing on top of the wall may be needed for safety. To accomplish this, for example, the masonry wall itself can be built higher by using screen block units.

Paving

Concrete masonry units are used for paving driveways, access lanes, parking areas, streets, plazas, shopping malls, walks, patios, and floors on grade, to name just a few applications. They are also used for slope paving under highway or railway grade-separation structures and on other steep embankments to prevent costly and often dangerous soil erosion, particularly where grass will not grow to protect the surface. Produced in a wide range of shapes and colors, paving units are easy to handle and install, requiring only a few tools.

Vehicular and Pedestrian Traffic

The concrete masonry industry offers a variety of interlocking paving units (also called pavers, concrete pavers, paving stones, paving block, and brick pavers) in interesting patterns, bright with color and human in scale (Fig. 8-26 through 8-43). They are used for foot traffic, light vehicle traffic, and special units are used for heavy traffic.

The units are not greater than 6½ in. wide, 9½ in. long and 5½ in. thick, and they have a compressive strength of at least 8,000 psi. Conventional concrete ingredients are used in their manufacture, however, to prevent scaling, fly ash should be avoided in units to be exposed to deicing chemicals. Paving units should meet the requirements of ASTM C936. Design procedures are available in PCA's *Structural Design of Pavements Surfaced with Concrete Paving Block ("Pavers")*, IS228P. However, structural design is only required for the more heavily-loaded industrial pavements.

Installation of paving units can begin in the spring as soon as the frost is gone and the ground is dry and firm. In the fall, installation can continue until frozen ground prevents proper compaction of the subgrade material. The final job will be only as stable as the subgrade.

Requiring careful preparation, the subgrade should be uniform, hard, free from foreign matter, and well drained. The best masonry paving installations are made by removing all organic matter such as grass, sod, and roots. Hard spots are loosened and tamped to provide the same uniform support as the rest of the subgrade. Mucky spots are dug out, filled with soil similar to the rest of the subgrade or with granular material (such as sand, gravel, crushed stone, or slag), and compacted thoroughly. All fill materials should be uniform, free of vegetable matter, large lumps, large stones, and frozen soil.

Fig. 8-26. Interlocking concrete paving units.

Fig. 8-27. Concrete pavers provide a durable and attractive walkway for this precast concrete structure. Construction of this walkway is illustrated in Figs. 8-28 through 8-42.

Paving units are usually bedded in sand or sometimes in a durable air-entrained mortar. Sand bedding gives excellent, long-lasting results and is ideally suited for residential, commercial, and industrial applications as well as for the do-it-yourselfer in such applications as walks, patios, and temporary pathways. The units are easily lifted for maintenance work or relocated if desired. Edge restraint such as precast or cast-in-place concrete curbs, steel and plastic restrainers, anchored timber, or existing structures usually is needed to prevent the units from creeping apart.

Detailed photographs illustrating the placement of concrete pavers are shown in Figs. 8-27 to 8-42.

Slope Paving

With ungrouted or grouted joints, concrete masonry is an economical and pleasing solution to the slope paving problem. Construction costs are low due to minimal materials-handling at the site and the ease of placing units on the slope (Fig. 8-44).

Masonry Units Required

Different sizes and shapes of paving units (Fig. 8-45) are used for slope paving, with unit thickness varying from 4 to 8 in. The thicker units are used for severe exposure such as on riverbanks. The slope paving units most often specified are standard 8×16-in. solid units* (Fig. 8-45a), although it is possible to manufacture paving units up to about 16×24 in. in size.

In areas where there are no freezing temperatures or if there is proper drainage, cored masonry units have been used successfully. They are lighter in weight and often less expensive to lay than the 100% solid units. However, the solid units are preferred in freeze-thaw climates because there is no chance of water freezing in core spaces. Also, 100% solid units discourage vandalism (by additional weight), especially in ungrouted installations.

---

*Solid units for paving have no voids (100% solid), whereas solid units for other applications may have up to 25% voids.

Fig. 8-28. Once the subgrade is excavated, leveled, and free of organic matter, soft spots, and foreign material, the edge restraint can be installed. Reinforced-concrete curbs (placed here) provide the greatest edge security for pavers.

Fig. 8-29. Crushed stone (¾ in. maximum size) is provided for a 5-in.-thick granular base over the subgrade. Also note that the concrete curbing is moist cured with wet burlap covered with plastic sheets.

Fig. 8-30. The crushed aggregate base is compacted to 95% density with a vibratory plate compactor.

Fig. 8-31. The base should be compacted by hand tampers in locations, such as corners, that are not accessible to mechanized vibrators.

Fig. 8-32. A fine aggregate bedding course is placed over the coarse aggregate base. Sand—the preferred bedding material—should meet the requirments of ASTM C33 for fine aggregate for concrete. Aggregate rakes are very helpful for distributing the sand.

Fig. 8-33. The sand is leveled to a thickness of 1½ in. A trowel is helpful with this screeding procedure. The sand bedding course is not compacted at this stage of construction. It is assumed that after the pavers are placed and vibrated, elevation of the sand will be approximately ½ in. lower than screeded level.

Fig. 8-34. Paving units are placed hand-tight directly on the leveled sand. Placement should start from a corner. Most paver designs include special edge units that can be used to greatly minimize cutting of units.

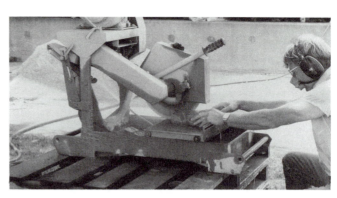

Fig. 8-37. Pavers can be cut with a wet table saw using diamond blades (pictured) or a masonry splitter (called a guillotine). The layout should be designed to avoid cutting of pavers to reduce installation costs, however, most jobs will require some cutting.

Fig. 8-35. On small jobs the units are usually placed one at a time by hand, although machines are available that can place 35 to 40 units at a time. For efficiency, a supply of units is placed near the mason. A rubber mallet is used to help set the pavers and keep the rows of pavers aligned.

Fig. 8-38. Along the edge restraint, special edge units can be installed or the units can be cut to the required dimension. The pavers should be placed snugly against the edge restraint. Cut units are shown being installed; special edge units are stacked next to mason.

Fig. 8-36. A stringline (bottom of photo) controls the paver elevation and was set ½ in. above the elevation of the edge restraints to allow for minor future settlement. The pavement should be sloped for drainage; this project was slightly crowned to provide side-to-side drainage.

Fig. 8-39. After the pavers are in place, dry sand is broomed onto the surface. A fine masonry sand is usually used because the small particles easily penetrate the joints between the pavers.

Fig. 8-40. The pavers are compacted into the loose bedding sand below with a vibratory compactor. The sand at the surface fills the joints and locks the units into place. This operation also corrects any slight height variations between pavers, leaving a smooth level surface. Many specifications also require one pass of the compactor prior to surface sanding. Sand is added and compaction continued until the joints are full. The compactor shown has rubber rollers to help prevent marring of the paver surface.

Fig. 8-41. After compaction, the excess sand is broomed off (*above*) and the surface rinsed clean (*below*).

Fig. 8-42. Completed pavement.

Any concrete masonry units selected for slope paving should have a minimum compressive strength of 3,000 psi on the gross cross-sectional area at the time of delivery to the jobsite.

Construction Features

For slope paving, some specifications limit the maximum angle of slope to 35 deg.; others specify a maximum slope commensurate with the angle of repose of the underlying material. Actually, ungrouted masonry paving units can be laid on any angle at which the underlying material can be stabilized.

A 2- to 4-in. layer of granular material—sand, crushed stone, or gravel—should be installed immediately below the masonry units to facilitate drainage and minimize the possibility of frost heave. When levelness between units is important, sand is used because it can be struck off very smoothly. The allowable surface variation is ¼ in. in 10 ft.

When slope paving units are not grouted together, they are laid tightly against each other. If grouting is required, ¾- or 1-in.-wide joints can be left for grout fill. Some contractors use wooden spacers to ensure this width, since spacing closer than ¾ in. makes it difficult to completely fill the joints with grout. Grout proportions, by volume, should be 1 part portland cement to 3 parts sand, with just enough water to make the grout workable. Interlocking units (Fig. 8-45) do not require grouted joints. An ungrouted installation offers several advantages:

1. If there is settlement or frost heave, the paving units can be adjusted individually.
2. If appearance allows, the underlayment need not be so carefully struck off and thus the use of crushed stone or gravel becomes more feasible.

Fig. 8-43. Interlocking concrete masonry pavers make an uncommonly attractive auto showroom floor.

Fig. 8-44. Slope paving units are laid under a highway bridge to prevent erosion.

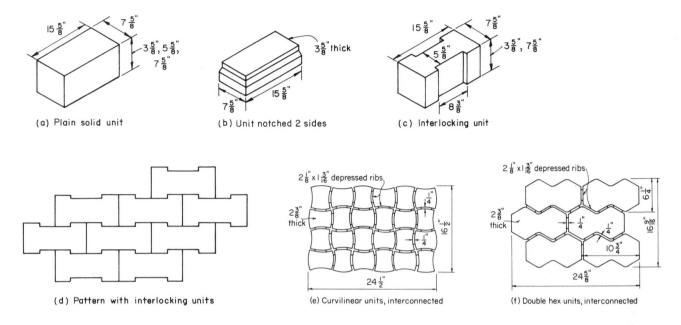

(a) Plain solid unit (b) Unit notched 2 sides (c) Interlocking unit

(d) Pattern with interlocking units (e) Curvilinear units, interconnected (f) Double hex units, interconnected

Fig. 8-45. Concrete masonry slope paving units.

3. Ungrouted construction is less expensive and the units can be easily replaced.

The advantages of grouted construction are:

1. Percolation of water between and under paving units is avoided.
2. There is less chance of undermining the slope protection.
3. The units are more securely held in place, deterring vandalism and theft.

Most concrete masonry slope paving installations are built with some type of support at the bottom or toe of the slope to prevent sliding of the units and to provide a straight, firm foundation for the first course of masonry. In some cases the support may be provided by a compact, level surface of embankment material. Toe construction will vary, depending on conditions of drainage, paving, and grade at the bottom of the slope (Fig. 8-46).

It should be noted that, instead of a large support at the bottom of the slope, smaller intermittent supports may be made on the slope itself. For example, each tenth course of masonry could be embedded into the slope, with the long dimension of the block perpendicular to the slope.

Drain troughs are often provided at the sides of the area paved with masonry (Fig. 8-47) and channelized waterways also carry water from the toe of the slope to a natural outlet. These waterways should be paved for a sufficient distance to eliminate any possibility of erosion undermining the slopes.

Where drains are not necessary, care should be exercised that the fill material at the edge of the paving

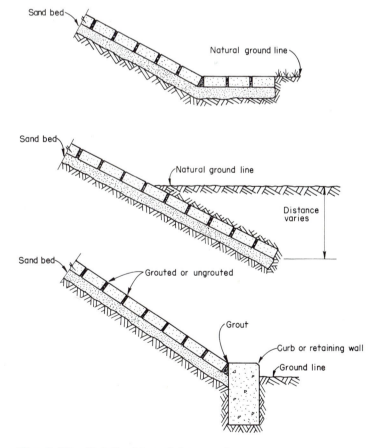

Fig. 8-46. Details at toe of slope paving.

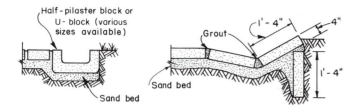

Fig. 8-47. Edge drains for slope paving.

units is level with the top of the units. This is especially important to hold the edge units in place when joints are not grouted.

Riverbank Revetments

As an alternative to stone riprap, special concrete revetment block are available for controlling erosion along the banks of rivers and lakes. The perforated waffle units shown in Fig. 8-48a may be laid on the bank beneath the water, one unit ajdjacent to the next. Grout is needed only on the perimeter of the revetment to keep these units in place, and construction costs can be lowered by inserting precast keys or stakes in the preformed key holes. The cores in the units may be filled with sand and gravel, and vegetation will grow through the cores, further securing the units in position. Such a revetment functions as an articulated mat, with each unit settling individually to a firm resting position.

The ribbed waffle units (Fig. 8-48b) used for erosion control are 90-lb. units that can be laid by unskilled volunteers during flood emergencies. Interlocking units (Fig. 8-45c) are also used for erosion control.

Other Paving

Waffle units (Fig. 8-48) are useful as turf block or "grass pavers," as pictured in Fig. 8-49. They are laid to create light pavement at access lanes and parking areas. Grass will grow up in the cores or between the ribs despite frequent traffic or parking of vehicles.

Patio block are commonly 8×16 in. (nominal), with thicknesses of 1⅝, 2¼, and 3⅝ in. Various shaped units are available in a natural grey, white (by use of white cement), or tones of red, black, brown, green, or yellow. Almost any pattern used for masonry walls may be used for patio block.

Fig. 8-49. Ribbed waffle units may be used for overflow parking areas. Grass grows between the ribs to enhance appearance.

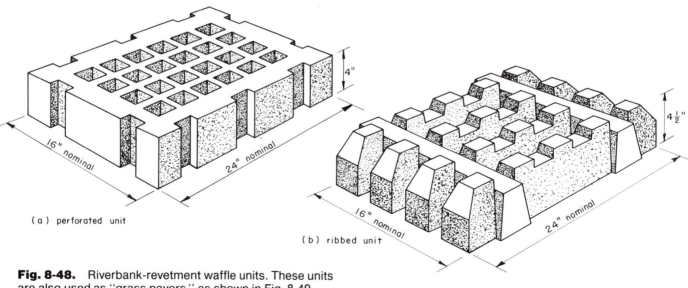

Fig. 8-48. Riverbank-revetment waffle units. These units are also used as "grass pavers," as shown in Fig. 8-49.

Catch Basins and Manholes

Concrete masonry construction is an accepted method for building catch basins, inlets, manholes, valve vaults, pump wells, and other shallow-depth, circular, underground structures. A typical catch basin and some sewer manholes are shown in Figs. 8-50 and 8-51, respectively.

Catch basins have space at the bottom for the settlement and storage of suspended solids that might otherwise be carried and deposited in the pipeline. If the catch basins are part of a sanitary sewerage system, they should be provided with solid covers to prevent sewer odors from reaching the street.

Drop manholes are constructed at intersecting lines or where there is an abrupt drop in elevation in a sewer line. The arrangment shown in Fig. 8-51c is desirable in that it reduces turbulence and prevents sewage from splashing on men working in the manhole. Such construction still permits cleaning of the sewer.

Materials

Concrete block should meet the absorption and strength requirements of Standard Specificaiton for Concrete Masonry Units for Construction of Catch Basins and Manholes, ASTM C139. This specification is for segmental masonry units. When concrete brick are used, they should be plastered on the outside with Type S mortar.

Batter block (Fig. 1-23, Chapter 1) are very useful for cone construction atop the barrel of a catch basin (Fig. 8-52) or manhole. The cone reduces the inside diameter to 2 ft. at the top to receive a standard manhole cover, as shown in Figs. 8-50 and 8-51. No cutting of units is necessary because block producers have predetermined the exact number and size of batter block required.

For a manhole cover size for which batter block are not available, the masonry wall of the structure is continued without reduction in diameter for its entire height. A precast or cast-in-place concrete slab is then added.

Special block are available to frame around inlet and outlet pipe.

Construction

For a catch basin or manhole, excavation should be to the required depth and dimensions specified. If rock is encountered, the bottom of the excavation should be carried down at least 6 in. below the elevation of the bottom of the structure and backfilled with sand.

The bottom of the structure may be constructed of cast-in-place concrete or a precast concrete slab, with the concrete having a water-cement ratio, by weight, of

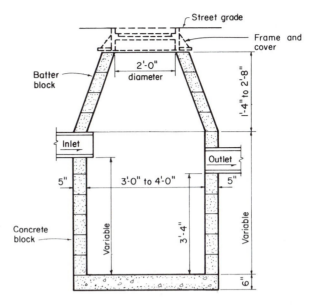

Fig. 8-50. A typical catch basin.

no more than 0.50 and a minimum compressive strength of 3,000 psi at 28 days. The use of a precast bottom is finding increasing favor because it minimizes the time the excavation must be kept open. It is of particualr value during wet weather or when the excavation is in wet subsoil.

The masonry wall of the structure is constructed in horizontal courses, with vertical joints staggered. All joints are completely filled with Type M mortar. Any castings that will be used should be set in a full bed of mortar. The masonry units around inlet or outlet pipe are carefully laid and sealed with mortar to prevent leakage. Special block are useful to frame around inlet and outlet pipe.

Heavily galvanized or other noncorrosive ladder rungs may be attached to a manhole wall or embedded in it on about 16-in. centers. Otherwise, a ladder may be installed in the manhole.

Granular material such as sand, gravel, or crushed stone is used to backfill the completed structure. Backfill material may be governed by specifications or approval of the engineer.

Storage Bins

Concrete masonry units are popular for construction of storage bins for grains, fruits, and vegetables. Where drying of the stored produce is not important, masonry units are generally laid in the conventional manner. Where produce drying is necessary, the units are laid with cores horizontal, as shown in Fig. 8-53.

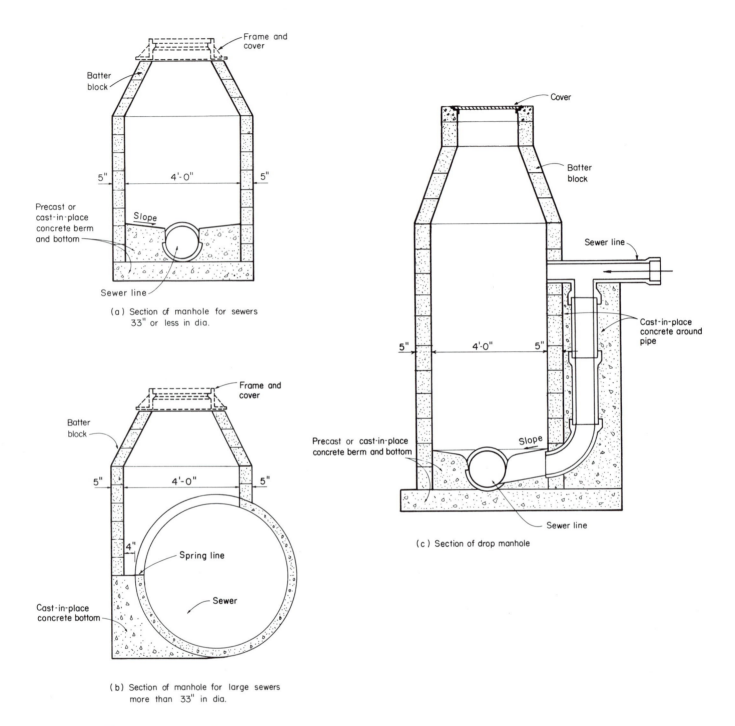

(a) Section of manhole for sewers 33" or less in dia.

(b) Section of manhole for large sewers more than 33" in dia.

(c) Section of drop manhole

Fig. 8-51. Types of sewer manholes.

Fig. 8-52. Concrete block catch basin in a storm sewer system.

Fig. 8-53. Constructed of concrete block laid with cores horizontal, this rectangular crib provides durable, safe storage for ear corn.

Details of construction for concrete masonry storage bins are given in Fig. 8-54. Pilasters for lateral support against the pressure of the stored produce may be built of reinforced concrete masonry or reinforced concrete. They may be placed outside the walls (Fig. 8-54) or flush (Fig. 8-53).

Protection against rodents is provided by installation of hardware cloth and metal joint strips, as shown in Fig. 8-55.

Prefabricated Panels

Prefabrication in the construction industry extends to concrete masonry where wall panels are preassembled away from the building or off the jobsite for speed, efficiency, and quality of construction. In one elementary method, wall panels are constructed on reinforced concrete grade beams and then trucked to the site and set in place with a crane (Fig. 8-56). In another method, the concrete masonry wall panel is constructed on a working slab. Lifting hooks are built into the panel, and the strength of the grout holds the panel together when it is lifted by the top. If the panel is to be moved to a storage area for curing, a steel bed frame is used for support. In addition to building construction, prefabricated panels have become popular in the construction of noise-barrier walls along busy streets and highways.

Hand-built wall panels (Fig. 8-58) may use mass-production refinements. For example, mortar proportioning and mixing can be controlled carefully, and electric scaffolds that rise with the work permit masons to work in the most comfortable and productive position.

Wall panels may also be built offsite by a machine that lays block at the rate of one every five seconds, much faster than a mason can work (Fig. 8-59). The machine completely fills the head joints as the concrete block is vibrated into place. Cross webs as well as face shells are mortared using this highly mechanized operation. The mortar is of very high quality; no admixtures are used.

The cost of prefabricated wall panels is comparable to the cost of walls that are laid in place. However, there are advantages of prefabrication that produce appreciable savings and other benefits. Not only are better quality and uniformity of construction more easily achieved, but also the speed of erection is quite distinct. Work can proceed on the panels in bad weather. Masons working in shop buildings or other enclosures are assured continuous work with little interruption. Also, cumbersome scaffolding and its winter enclosures are eliminated, and the accident rate among masons in prefabrication work is lower.

Prefabricated panels should meet the requirements of ASTM C901, Standard Specification for Prefabricated Masonry Panels.

Foundation Block

Foundation block are concrete masonry units used to replace cast-in-place concrete footings. A complete

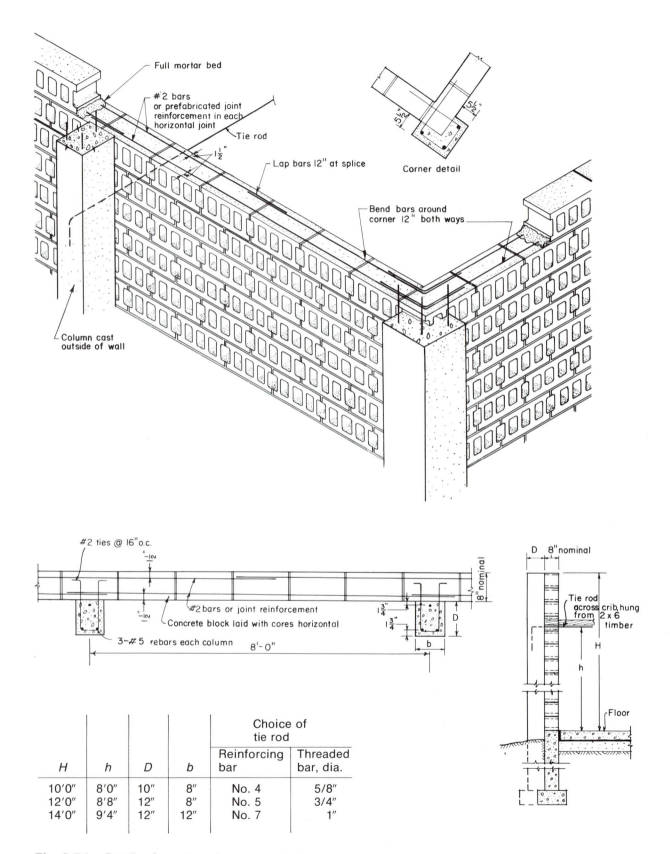

Fig. 8-54. Details of a rectangular concrete masonry corn crib.

H	h	D	b	Reinforcing bar	Threaded bar, dia.
10'0"	8'0"	10"	8"	No. 4	5/8"
12'0"	8'8"	12"	8"	No. 5	3/4"
14'0"	9'4"	12"	12"	No. 7	1"

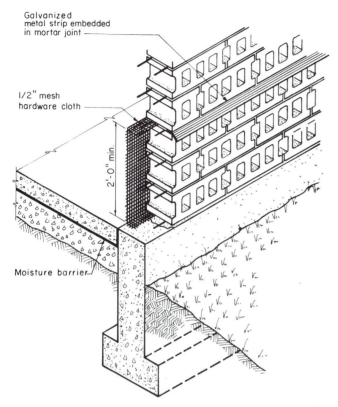

Fig. 8-55. Installation of metal strip shield and hardware cloth to keep rodents out of a concrete masonry corn crib.

Fig. 8-57. Prefabricated wall panels shown braced into position until the framing above is completed.

Fig. 8-58. Masonry panels are fabricated at the site in a canvas-covered shelter, where they are stored for 7 days before being lifted by crane onto the building.

Fig. 8-56. Panel of concrete masonry preassembled for residential construction.

Fig. 8-59. Plant fabrication of panels. Concrete block are conveyed by belt to a machine that automatically assembles them into wall panels. The finished panels are rolled along a track to a storage area.

foundation system of interlocking units can be used for 1- and 2-story residential, commercial, and agricultural structures. Homeowners can also use the units as foundations for garages, house additions, sunrooms, greenhouses, and storage buildings. In one proprietary system for 8-in. walls, footer unit dimensions are 4×8×16-in. with a unit weight of about 24 lb. (Fig. 8-60). The units are placed 12 in. below the frost line directly on undisturbed soil that can be finished and leveled with crushed stone, gravel, mortar, or lean concrete. The block foundation should meet the requirements of applicable building codes for masonry footings. For more information, see References 72 and 93.

Imitation Stone

Imitation-stone concrete masonry units are available in a multitude of colors and shapes to provide a natural stone-like appearance to residential, commercial, and industrial structures. The units can be made to resemble cobblestone, limestone, granite, lava, and other stones in ledge or boulder fashion (Fig. 8-61). The units are usually made with lightweight aggregate to reduce the weight of the units for ease in handling and to reduce the dead load of a structure. For more information, see Reference 85.

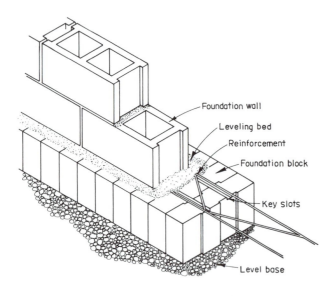

Fig. 8-60. A foundation system using a 4×8×16-in. interlocking concrete masonry footer block.

Fig. 8-61. Ledge rock type imitation stone (Reference 85).

Chapter 9

MAINTENANCE OF MASONRY STRUCTURES

Maintenance is an important part of all structures. The useful life of a building can be directly related to the quality of the maintenance. Without preventative maintenance and prompt repair when needed, costly damage can be incurred by the building and its contents. A good maintenance program will greatly reduce the chances of major problems in a structure. The cost of repair of major damage usually far exceeds the cost of proper and timely maintenance. Concrete masonry is a very durable construction material and usually requires very little maintenance. Maintenance problems that do occur are often the result of poor design or construction practices. This chapter reviews common maintenance and repair issues.

Inspection

Masonry structures should be periodically inspected for any signs of unit deterioration, water leakage, joint deterioration, or other problems. For example, cracking can be a sign of structural problems or simple

Fig. 9-1. Fiber-optic boroscope used to analyze the condition within a concrete masonry wall.

settlement; without repair, the crack can result in accelerated deterioration of a wall and excessive water leakage. Most movement joint sealants have a limited life and must be replaced periodically.

Inhouse inspections should be performed annually and an inspection by a masonry or building specialist should occur every five years. The owner of a structure should have a copy of all building drawings, material specificiations, a list of suppliers of products and materials used in the structure, test reports, and other applicable construction documents.

The inspection should identify areas in the building that may have potential problems and these areas should be monitored over time. The inspector should compare protected parts of a structure with weather exposed areas to monitor deterioration or problem development due to weathering.

The inspection will first reveal if a problem exists. The extent and cause of the problem is then determined. A problem in one part of the masonry may not be obvious, but the results of the problem can appear in another part of the structure. For example, damaged flashing may not be apparent to the casual observer at the flashing location, but water stains several feet away are readily recognizable. Once the location and cause of the problem are known, a repair procedure can be designed and implemented. After the repair is complete, the area should be monitored for the development of similar problems.

Usually a visual inspection will detect most problems. Special equipment such as the boroscope in Fig. 9-1 is available to help analyze problems. Nondestructive and destructive testing are also helpful.

Weathering and Frost Damage

Weathering and frost damage of masonry is characterized by spalls, cracks, and surface erosion. Masonry building elements exposed on both sides, such as parapet walls, screen walls, fences, fully exposed columns, and free standing walls, are especially vulnerable to weathering. Spalled units near shelf angles can indicate structural problems. Masonry beneath recesses or projections or in locations prone to collect water should be closely monitored, especially if flashing is not present. Water must not be allowed to accumulate in masonry cells to prevent the disruptive force of ice formation. Water expands about 9% upon freezing and can exert thousands of pounds of force on a unit, far exceeding its strength. Severely deteriorated masonry units can be replaced. For extreme conditions involving an entire wall, one solution may be to apply exterior insulation and portland cement plaster or other finish system (see "Exterior Insulation and Finish Systems" in Chapter 7).

Mortar must also be inspected for weathering and frost damage. Deteriorated mortar often expands, interrupting the alignment of units. Sometimes mortar deteriorates to such a degree that it falls out of the joint. The condition of mortar can be assessed by scratching the mortar surface with a knife or nail. A slight amount of dusting of old mortar is common and usually not a problem. Separation of mortar from the unit can also occur due to incompatibility between mortar and unit (due to absorption of the unit), bulging of the wall, and other reasons. Mortar can be replaced by tuckpointing, discussed later in this chapter.

Cracks and Spalls

Cracks, spalls, and wall deflection or bulging are all signs of uncontrolled or unaccomodated movement. Wall deflection or bulging can be the result of ice formation in a cavity, insufficient anchoring with ties, rusted ties, or structural failure. Cracks and spalls often accompany masonry deflection or bulging. Proper construction and spacing of control joints should accomodate movement resulting from moisture and temperature changes. Masonry building elements exposed to weather on both sides experience more thermal and moisture induced movement than nonexposed masonry and therefore tend to have more cracks. A lack of proper jointing can be corrected by cutting new joints into a wall. Sometimes a joint does not perform properly because it is inadvertently bridged with mortar. This condition is easily corrected by cleaning the hardened mortar from the joint. Other conditions that restrain joint movement should also be corrected.

Cracks can occur where a change in materials occurs. Vertical cracks form in the center of walls that are overloaded. Cracks occur around improperly designed openings in masonry as well as near corners that are twisting due to structural restraint. Bond-breaking flashing should be provided between masonry and the foundation to prevent spalls and cracks at foundation corners.

Cracks form upon extensive rusting of lintels and shelf angles. In repairing shelf angles, stress should be relieved from the top down to avoid damage to headers or transverse bond within mortar joints. Stress relief is provided by cutting new movement joints or relieving restraint at existing movement joints.

Wall deflection or bulging due to a lack of ties can be corrected by installing special repair anchors or ties that are properly spaced throughout the wall. Bulged masonry must be straightened and reanchored prior to other repairs. Spalled or cracked units can be replaced and damaged mortar joints can be tuckpointed. The degree of repair depends on the exposure condition and the amount of damage incurred.

Sealants

Movement joint sealants can deteriorate and crack over time due to weather exposure, especially drying over time, ultra-violet sunlight exposure, and movement fatigue. The sealant may break away from the masonry (adhesion failure) or tear within the sealant (cohesion failure). Without replacement, rain can enter the joint and cause water damage within the building, rust lintels and shelf angles, or simply add to efflorescence or freeze-thaw damage of masonry units. The old, deteriorated sealant must be completely removed and the joint cleaned and primed before installation of the new sealant. Four-in. grinders are available for removing sealants, however, they do not work well because the high speed of the cutting wheel generates heat and melts the existing sealant. A vibratory tool, a knife of sorts, is available to remove sealants. The knife blade is inserted at the interface between the sealant and the masonry, and the sealant is essentially severed from the masonry.

Movement joints must be properly spaced to prevent over extension or compression of sealants. The type of sealant, the shape of sealant (width and depth), and the amount of restraint dictate its performance upon deformation. The center depth of the sealant should be about one-half the sealant width.

Adhesive failure between the sealant and the masonry usually indicates poor surface preparation prior to applying the sealant. Porous surfaces should be primed and closed-cell backer rod should be compressed about 50% during installation. A bond breaker should be used to prevent adhesion at the bottom of the sealant. Tearing apart of the sealant (cohesive failure within the sealant) occurs when the wrong sealant is used, the sealant shape is poor, or the joints are too far apart, putting excessive stress on the sealant.

Water and Air Permeation

Water and air can permeate a wall through failed joint sealants, through cracks in mortar or masonry units, or through areas around improperly placed flashing or where there is a lack of flashing. (Weepholes are important to proper drainage.) Efflorescence on the masonry or puddles on the floor in a building are usually evidence that water has permeated a wall.

Air moving in and out through cracks and openings can convey moisture through masonry walls. Upon exposure to cool temperatures, the moisture can condense into a liquid in or on the wall. In high-rise buildings, air leakage is aggravated by the "stack effect"—the inward (at the bottom) and outward (at the top) movement of air due to pressure differences generated by temperature and density variance between inside and outside air.

Patterns of efflorescence or dampness (sometimes frost in winter) seen on the masonry near the tops of buildings at certain times of the year are evidence of the outward flow of air. A crack between the masonry wall and a column, slab, or crosswall is an obvious path for air leakage, as well as where window frames, door frames, and ducts penetrate the wall. A joint sealant can stop this leakage effectively.

A continuous vapor barrier installed on the warm side of the wall can prevent the accumulation of a damaging level of moisture in the masonry units.

The presence of moist air or water in a wall can rust ties, lintels, and other metal components, resulting in cracks and wall deflection or bulging. Frost damage can also occur within the units and mortar when frozen wet. For water to enter a wall, rain must be present on the wall surface, cracks or other openings must be present, and the rain must be driven or drawn into the wall by wind, gravity or differential air pressure. To prevent water permeation, through-the-wall flashing should be used, including locations below caps and sills. Flashing should not stop short of the exterior part of the unit. Water should drain past the surface of the masonry and not drip into the unit. Drips are helpful in allowing water to fall away from the wall surface. Sills should be sloped to drain water away from the wall. Sills should also extend beyond the wall surface and include a drip on the underside.

Unless repaired promptly, water leakage and water accumulation in walls can induce severe problems in masonry structures. The location of water leakage can be found by spraying water on the masonry with a hose and observing any leakage. Properly designed cavity walls with flashing provide excellent resistance to water permeation. Breathable, water-repelling sealers are helpful in reducing water leakage, however, asphaltic coatings should be avoided as they can trap moisture in the wall.

Efflorescence

Efflorescence is a deposit, usually white in color, that may develop on the surface of masonry. Often it appears just after a structure is completed. Although unattractive and generally harmless, efflorescence deposits can occur within the surface pores of the material, causing expansion that may disrupt the surface.

Causes

A combination of circumstances causes efflorescence. First, there must be soluble salts in the material. Second, there must be moisture to dissolve the soluble salts. Third, evaporation or hydrostatic pressure must cause the solution to move toward the surface. And fourth, the solution must evaporate to leave salts behind as efflorescence.

All masonry and concrete materials are susceptible to efflorescence. Water-soluble salts that appear in chemical analyses as only a few tenths of a percent are sufficient to cause efflorescence when leached out and concentrated at the surface. The amount and character of the deposits vary according to the nature of the soluble materials and atmospheric conditions.

Efflorescence is particularly affected by temperature, humidity, and wind. In the summer, even after long rainy periods, moisture evaporates so quickly that comparatively small amounts of salt are brought to the surface. Usually efflorescence is more common in the winter when a slower rate of evaporation allows migration of salts to the surface. With the passage of time, efflorescence becomes lighter and less extensive unless there is an external source of salt. Light-colored surfaces show the deposits much less than darker surfaces.

Efflorescence-producing salts are usually carbonates of calcium, potassium, and sodium; sulfates of sodium potassium, magnesium, calcium, and iron (ferrous); bicarbonate of sodium; or silicate of sodium. However, almost any soluble salt that finds its way into masonry materials may appear as efflorescence; consequently, chlorides, nitrates, and salts of vanadium, chromium, and molybdenum occasionally cause efflorescence.

Fig. 9-2. Severe efflorescence on a concrete masonry wall.

Chloride salts are highly soluble in water, so the first rain often will wash them off.

In most cases, salts that cause efflorescence come from beneath the surface, but chemicals in the materials can react with chemicals in the atmosphere to form efflorescence. For example, in concrete masonry, mortar, or stucco, hydrated portland cement contains some calcium hydroxide as an inevitable product of the reaction between cement or lime and water. Calcium hydroxide brought to the surface by water combines with carbon dioxide in the air to form calcium carbonate, which appears as a whitish deposit.

Another source of salts is soil in contact with basement and retaining walls. If the walls are not protected with a good moisture barrier, the salts may migrate a foot or two above grade.

Since many factors influence the formation of efflorescence, it is difficult to predict if and when it will appear. However, efflorescence will not occur if

- soluble salts are eliminated
- moisture is eliminated
- water passage through the mass is prevented

Eliminating the Salts

In the selection of materials, all ingredients should be considered for their soluble-salt content. To reduce or eliminate efflorescence-producing soluble salts:

1. Never use unwashed sand. Use sand that meets the requirements of ASTM C33 for concrete, or ASTM C144 for mortar.
2. Use low-alkali cement.
3. Use dehydrated lime free from calcium sulfate when using lime for mortar or stucco.
4. Use clean mixing water free from harmful amounts of acids, alkalies, organic material, minerals, and salts. Do not use drinking water that contains sufficient quantities of dissolved minerals and salts to adversely affect the resulting construction. Do not use seawater.
5. Never use masonry units known to effloresce while stockpiled. Use only masonry units of established reliability. Use brick passing efflorescence tests in ASTM C 67.
6. Use insulating material free of harmful salts when walls of hollow masonry units are to be insulated by filling the cores.
7. Be certain that mixer, mortar box, mortarboards, and tools are not contaminated or corroded. Never deice this equipment with salt or antifreeze material.
8. Consider using autoclaved concrete masonry units.

Water-repellent surface treatments such as silicones decrease surface efflorescence by causing the dissolved salts to be deposited beneath the treated surface. However, localized accumulation of salts and their crystallization beneath the treated surface may cause surface spalling or flaking. When there are large amounts of salt in the construction material, use of a surface treatment may cause problems.

Eliminating Moisture and Water Passage

Low absorption of moisture is the best assurance against efflorescence. When masonry walls are constructed in accordance with recommendations in available literature, it is unlikely that water penetration will be a problem. Research shows that composite walls of brick and block are capable of resisting water infiltration regardless of the type and composition of the mortar used in laying up the walls (see References 94-96). Design and workmanship affect water permeance far more than materials do.

To eliminate moisture or moisture passage through the structure, these steps are recommended:

1. Prevent inadequate hydration of cementitious materials caused by cold temperatures, premature drying, or improper use of admixtures.
2. Prevent entry of water by giving proper attention to design details for correct installation of waterstops, flashing, and copings.
3. Cover the top course of masonry at the completion of each day's work, particularly when rain is expected.
4. Cure or dry concrete masonry units in the presence of carbon dioxide gas. This appears to be beneficial in changing calcium hydroxide to calcium carbonate, which seems to form in the pores at or just below the surface. The pores are thus partially filled, reducing the passage of water.
5. Install vapor barriers in exterior walls or apply vaporproof paint to interior surfaces and use designs that minimize condensation within masonry.
6. Apply paint or other proven protective treatment to the outside surfaces of porous masonry units.
7. Tool all mortar joints with a V- or concave-shaped jointer to compact the mortar at the exposed surface and create a tight bond between mortar and masonry units. Weeping, raked, and untooled stuck joints are not recommended except in dry climates. Deteriorated mortar joints should be tuckpointed to keep moisture out of the wall.
8. Carefully plan the installation of lawn sprinklers or any other water source so that walls are not subjected to unnecessary wetting.
9. If architecturally feasible, use wide overhanging roofs to protect walls from rainfall.
10. Design for pressure equalization between the outside and the void within the masonry wall.

How to Remove Efflorescence

Where there is efflorescence, the source of moisture should be determined and corrective measures taken to keep water out of the structure.

Most efflorescence can be removed by dry brushing, water rinsing with brushing, light waterblasting, or light sandblasting followed by flushing with clean water (see ASTM D4261). If this is not satisfactory, it may be necessary to wash the surface with a dilute solution of muriatic acid (1 to 10 percent).* For integrally colored concrete masonry, only a 1 to 2 percent solution should be used to prevent surface etching that may reveal the aggregate and hence change color and texture. A solution of 1 part vinegar to 5 parts water is also helpful.

Before applying an acid solution, always dampen the wall surface with clean water to prevent the acid from being absorbed deeply into the wall where damage may occur. Application should be to small areas of not more than 4 sq.ft. at a time, with a delay of about 5 minutes before scouring off the salt deposit with a stiff bristle brush. After this treatment, the surface should be immediately and thoroughly flushed with clean water to remove all traces of acid. If the surface is to be painted, it should be thoroughly flushed with water and allowed to dry. Refer to ASTM D4260 for more information.

It is often helpful to determine the type of salt in the efflorescence so that a cleaning solution can be found that readily dissolves the efflorescence without adversely affecting the masonry. Before any treatment is used on any masonry wall, the method should be tested on a small, inconspicuous area to be certain there is no adverse effect.

Since acid and other treatments may slightly change the appearance, the entire wall should be treated to avoid discoloration or mottled effects.

Tuckpointing

Tuckpointing is the act of replacing cutout or defective mortar joints with new mortar. There are two basic reasons why tuckpointing may be necessary: (1) leaks in the mortar joints and (2) deterioration of joints. Tuckpointing will produce a weathertight wall and help to preserve the structural integrity and the appearance of the masonry.

If a wall is being tuckpointed to make it weathertight, it is recommended that all mortar joints be tuckpointed. Minute cracks that could pass a visual inspection might still allow moisture to pass through

---

*Caution: Rubber gloves, glasses, and other protective clothing should be worn by workers using an acid solution. All precautions on labels should be observed because muriatic acid can affect eyes, skin, and breathing.

Table 9-1. Tuckpointing Mortar (Vertical Surfaces)

| Type of service | Mortar type (ASTM C270)* | |
	Recommended	Alternate
Interior	O	N or K**
Exterior above grade exposed on one side. Not frozen when saturated and not exposed to high wind or significant lateral load	O	N
Exterior, other than above	N	O† or mortar type based on structural or durability concerns

*Structural concerns dictating mortar type supercede these mortar recommendations. For pavements, mortar Types M or S with applicable frost resistance should be considered. Mortar should not contain more than one air-entraining material as high air contents reduce bond and compressive strength, but some air entrainment does improve freeze-thaw durability.

**Type K mortar consists of 1 part portland cement, 2½ to 4 parts hydrated lime, and sand is 2¼ to 3 times the sum of the volumes of cement and lime.

†Type O mortar is recommended for use where the masonry is aboveground and is unlikely to be frozen when saturated or unlikely to be subjected to high winds or other significant lateral loads. Type N or other appropriate mortar should be used in other cases.

Adapted from ASTM C270.

the masonry. Before the start of any tuckpointing intended to produce weathertight masonry, a thorough inspection of all flashings, lintels, sills, and calked joints should be made. This is to ensure that water will not leak through into the masonry.

If it is obvious that the water is leaking through only one crack, it may be sufficient to tuckpoint only the mortar joints in the vicinity of the crack.

Preparation of Joints

Mortar joints should be cut out to a depth of at least ½ in. and in all cases the depth of mortar removed should be at least as great as the thickness of the mortar joint (see Fig. 9-3). If the mortar is unsound, the joint should be cut deeper until only sound material remains. Shallow (Fig. 9-4) or furrow-shaped (Fig. 9-5) joints will result in poor tuckpointing.

A tuckpointer's grinder with an abrasive blade is usually more efficient than hand-chiseling for cutting out defective mortar. All loose material must be removed with an air jet or a water stream.

The joints between clay masonry units should be dampened to prevent absorption of water from the freshly placed mortar. However, the joints should not be saturated just prior to tuckpointing because free water on the joint surfaces will act to impair bond. To avoid shrinkage, the joints between concrete masonry

units should not be wetted before or during tuckpointing.

Mortar Ingredients

Foremost among the factors that contribute to good tuckpointing mortar is the quality of the mortar ingredients. The following material specifications of ASTM are applicable:

Masonry cement—ASTM C91 (Type N)

Portland cement—ASTM C150 (Types I, IA, II, IIA, III, or IIIA)

Blended hydraulic cement—ASTM C595 [Types IS, IS-A, IP, IP-A, I(PM), or I(PM)-A]

Hydrated lime for masonry purposes—ASTM C207 (Types S, SA, N, or NA)*

Quicklime for structural uses (for making lime putty)—ASTM C5

Sand—ASTM C144

For mortar joints that are less than the conventional ⅜ in. thick, 100% of the sand should pass the No. 8 sieve and 95% the No. 16 sieve.

If only portions of a wall are tuckpointed, the color and texture of new mortar should closely match the old mortar by careful selection and proportioning of mortar ingredients. Admixtures should not be used unless specified.

Preparation of Mortar

Tuckpointing mortar should have a compressive strength equal to or less than that of the original mortar or should contain approximately the same proportions of ingredients as the original mortar. Recommended mortar types are shown in Table 9-1. Some masonry applications with structural concerns or severe frost or environmental conditions (such as horizontal surfaces exposed to weathering) may require the use of special mortars other than those in Table 9-1.

A recommended procedure for mixing tuckpointing mortar is (1) mix all of the dry material, thoroughly blending the ingredients; (2) mix in about half the water or enough water to produce a damp mix which will retain its shape when formed into a ball by hand; (3) mix the mortar for 3 to 7 minutes, preferably with a mechanical mixer; (4) allow the mortar to stand for one hour (but not more than 1½ hours) for prehydration of the cementitious materials to reduce shrinkage; and then (5) add the remaining water and mix for 3 to 5 minutes.

*Types N and NA lime may be used only if tests or performance records show that these limes are not detrimental to the soundness of mortar.

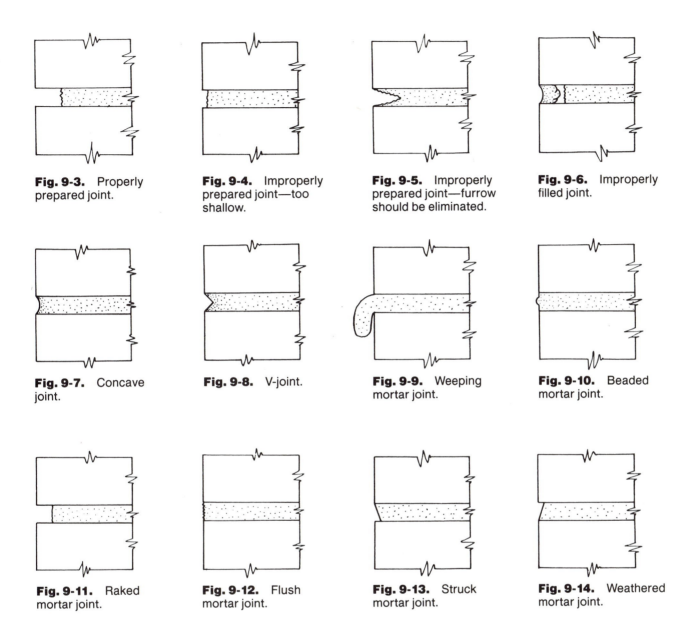

Fig. 9-3. Properly prepared joint.

Fig. 9-4. Improperly prepared joint—too shallow.

Fig. 9-5. Improperly prepared joint—furrow should be eliminated.

Fig. 9-6. Improperly filled joint.

Fig. 9-7. Concave joint.

Fig. 9-8. V-joint.

Fig. 9-9. Weeping mortar joint.

Fig. 9-10. Beaded mortar joint.

Fig. 9-11. Raked mortar joint.

Fig. 9-12. Flush mortar joint.

Fig. 9-13. Struck mortar joint.

Fig. 9-14. Weathered mortar joint.

Tuckpointing mortar should have a drier consistency than conventional mortar for laying masonry units. Evaporation and absorption may require that water be added and the mortar remixed to regain proper workability. Retemper as needed; however, the mortar should be discarded 2½ hours after the initial addition of water to the mix. Colored mortars may lighten upon the addition of water; therefore, retempering of colored mortar should be avoided.

Filling the Mortar Joints

The general method of applying mortar in joints that are to be tuckpointed is to use a hawk and a tuckpointing trowel. The hawk is used to hold a supply of mortar; it also catches mortar droppings if held immediately adjacent to the wall just below the joint that is being filled.

The tuckpointing trowel should be narrower than the mortar joints that are being filled in order to obtain a proper degree of compaction. If the trowel does not fit into the joints it will be more difficult to obtain thoroughly compacted and completely filled joints.

Mortar should be spread into a joint in layers and firmly pressed to form a fully packed joint. Firm compaction is necessary to prevent voids as shown in Fig. 9-6. The act of firmly compacting the mortar also helps ensure bond to masonry units and to the old mortar. Voids are undesirable because they may trap water which can freeze and damage the new joint.

Tooling Mortar Joints

Tooling compacts the mortar to a dense surface with good durability. For weathertight construction, all mortar joints should be tooled to a concave or V shape (see Figs. 9-7 and 9-8). These shapes are recommended because they do not allow water to rest on the joint and they result in the mortar being pressed toward both the lower and upper masonry unit. This reduces weathering and helps ensure maximum bond between the mortar and the masonry units.

Jointing tools can be made from round or square bar stock. For horizontal joints, or vertical joints in stacked bond (a bond pattern in which the joints form a continuous vertical line down the wall), the tool should be considerably longer than the masonry units to avoid a wavy joint.

Other Types of Mortar Joints

Weeping, beaded, raked, flush, struck, and weathered joints (see Figs. 9-9 to 9-14) can be used for decorative effects but are not recommended for maximum weathertightness and durability. They can be satisfactorily used for interior work and for exterior work when the masonry is protected from the elements or located in climates that do not impose extremely low temperature or water saturation.

Curing

The cementitious materials in mortars require a period in the presence of moisture to develop proper strength. The mixing water in the mortar will usually provide this necessary moisture. However, freshly placed mortar should be protected from the sun and drying winds. With severe drying conditions, it may be necessary to cover the masonry structure with polyethylene sheeting or use a fine water-fog spray on clay masonry for about 4 days to reduce evaporation of water from the mortar.

Cleaning Concrete Masonry Surfaces

The method selected for cleaning concrete masonry depends on the purpose of the cleaning and on the extent of the work to be done. It may entail a bucket and brush, a hammer and chisel, water-pressure and steam washing, grit blasting, chemical cleaning, or special mechanical power tools.

When faced with the decision to clean concrete masonry, a careful investigation is advisable, which may bring to light unexpected facts. Accurate diagnosis of the problem is essential for effective and successful cleaning. There is no really simple technique for normal situations: Water and chemical cleaners can lead to other problems caused by excessive moisture or unanticipated chemical reactions; grit blasting will change the texture and appearance of the surface; power tools can damage thin sections or remove more material than is desirable. Oils, grease, and certain penetrating chemicals must sometimes be removed before water or abrasive cleaning methods are used.

When cleaning concrete masonry is necessary, the following text will provide some guidance for selecting the least damaging method. The methods described below have merit for removing stains for appearance purposes, preparing surfaces for coatings and sealers, and preparing concrete masonry for repair work or plaster applications. For stain removal and some coating applications, it is desirable to minimize abrading of the concrete surface; whereas an abraded, rough texture is desired for repair work, plastering, and certain coating applications. For more information, see ASTM D4258, D4259, D4260, D4261, D4262, D4263, and PCA's publication, *Removing Stains and Cleaning Concrete Surfaces,* IS214T.

Before deciding on a particular method, clean a relatively small, inconspicuous area to assess the efficiency of the method and the appearance and condition of the surface after the treatment. The reasons for cleaning must be considered carefully because results with methods intended to improve only the appearance can differ substantially from results with methods to prepare the surface for coating.

Abrasive Blasting

Dry (Sandblasting)

Dry abrasive blasting, such as sandblasting, drives an abrasive grit at concrete masonry surfaces to erode away dirt, paint, various coatings or contaminants, and any deteriorated or damaged concrete. Also see "Metallic-Shot Blasting."

Sandblasting changes the appearance of the masonry surface. It is left with a rougher texture that may hold even more dirt and pollutants than before and hasten the need for recleaning. Sandblasting removes the edges at arrises and sharp detail on masonry units. Sandblasting can provide an excellent rough-textured surface for bonded repair work, such as plastering.

Although the sandblasting operation is not complicated, certain procedures and precautions known to experienced operators should be followed to ensure a uniformly clean surface.

Dry abrasive blasting should be used only when other, usually less abrasive, techniques are not successful. Soft abrasives, such as nut shells, are less destructive than sand.

Sandblasting equipment is available in various capacities. A venturi-type nozzle should be used on the gun for its solid-blast pattern rather than a straight-bore nozzle that produces lighter fringe areas. A remote control system attached to the sandblast pot gives the operator instant control of starts and stops as well as direction. The man operating the gun must be protected from dust and rebounding grit by a well-fitting, air-line hood in which a positive pressure of clean, filtered air is maintained. Other members of the blasting team should wear suitable protective clothing and equipment such as an approved respirator under a hood. Silica dust is a particularly dangerous substance because free silica can cause lung damage. The grit and dust particles must be removed by air blasting, brooming, pressurized water, or vacuum methods before a coating, sealer, or plaster is applied.

Wet Abrasive Blasting

Wet abrasive blasting is very similar to dry abrasive blasting (sandblasting) except that water is introduced into the air-grit stream at the nozzle. An adapter is secured to the nozzle for attaching the water supply. The water eliminates most of the visible dust but smaller, harmful particles remain a hazard to health and the same protective equipment and clothing are needed as for dry abrasive blasting. The wet-abrasive-blasting method will avoid the nuisance of dust but it involves an extra operation of rinsing off the surface after blasting to remove residual dust and dirt scum.

In the wet-abrasive-blasting method, the water cushions the abrasiveness of the grit and therefore is less destructive than dry sandblasting. Wet abrasive blasting is also used on soft brick and stone and ornamental masonry where it is desireable to avoid damaging sharp details in the masonry units. Friable aggregate (without silica) and water can be combined and applied at low pressure to provide a scouring action without harming the masonry.

Metallic-Shot Blasting

Self-contained, airless, portable blasting equipment can effectively clean horizontal or inclined surfaces. This equipment is primarily helpful with cleaning concrete pavers, although special units are available for vertical work. The removal of surface contaminants such as old paint, dirt, and loose and weakened concrete is accomplished by the impact of metallic abrasives thrown by a rapidly rotating centrifugal wheel onto the surface to be cleaned. The equipment includes components for dust and noise control as well as the recovery, cleaning, and recycling of the metallic abrasive.

After the abrasive impacts the surface, it is passed through an air-wash separator that removes foreign materials; the recovered abrasive is recirculated through the blast wheel. Pulverized concrete, dust, and contaminants are removed to a filter-bag dust collector, making the method virtually pollution free.

Chemical Cleaning

The materials used in chemical cleaning can be highly corrosive and frequently toxic. They require special equipment for their application and protective clothing for workers. In addition, protection may be necessary for adjacent areas, nearby buildings, and lawns, trees, and shrubs. For these reasons, chemical cleaning is best left to the specialist. If, however, a novice undertakes the cleaning job, the directions that come with the cleaner should be carefully followed.

Chemical cleaners are often water-based mixtures formulated for use on specific types of concrete and masonry. Most of them contain organic compounds called surfactants (surface-active agents) that work as detergents to allow the water to penetrate the surface dirt or stain more readily, thus hastening its removal. In addition, the mixtures contain a small amount of either acid or alkali, which assists in separating the dirt from the surface. Solvent-based (nonwater) cleaners are also used.

Cleaning with proprietary compounds, detergents, or soap solutions generally requires the same procedure as given here for acid cleaning.

Acid cleaning is often suggested as a satisfactory method for cleaning a masonry surface. Hydrochloric acid, also known as muriatic acid, is widely used because of its ready availability. Hydrochloric acid should not be used in areas where chlorides are prohibited.

The procedure for cleaning concrete masonry using a diluted acid solution is as follows:

1. Mix a 10% solution of muriatic acid (1 part acid to 9 parts clean water) in a nonmetallic container. Pour the acid into the water to mix. Stronger acid solutions may have to be used if the cleaning action is insufficient.
2. Mask or otherwise protect windows, doors, ornamental trim, and metal, glass, wood, and stone surfaces from acid solutions.
3. Remove dust and dirt from the area to be cleaned and presoak or saturate with water.
4. Apply the acid solution to the damp surface with spray equipment, plastic sprinkling cans, or a long-handled stiff-fiber brush. Allow the solution to remain for 5 to 10 minutes. Nonmetallic tools may be used to remove stubborn particles.
5. Rinse thoroughly. Flush the surfaces with large amounts of clean water before they can dry. Acid solutions lose their strength quickly once they are in contact with cement paste in concrete masonry or mortar; however, even weak, residual solutions can be harmful. Failure to completely rinse the acid solution off the surface may result in efflorescence or other damaging effects. Test with pH paper and continue rinsing until a pH of 7 or higher is obtained (see ASTM D4260 and D4262 for more information).

Steam Cleaning

In steam cleaning, water is pumped to a flash boiler where it is converted to steam and then directed onto the concrete masonry. The use of stiff-bristle brushes usually is necessary to assist in removing dirt. Today, improved methods and cleaning products have largely supplanted steam cleaning for buildings, although steam can sometimes help remove deep-seated soiling after acid cleaning and reach awkward areas. Steam cleaning essentially leaves the concrete masonry surface intact.

Water Spray

Low Pressure

In low-pressure water spraying, only enough water is sprayed onto the surface to keep the deposits of dirt moist until they soften. Larger amounts of water are no more effective, and they might oversaturate a wall and penetrate to the building interior, causing additional problems. Cleaning should begin at the top of the structure so that surplus water will run down and presoften the dirt below. How long it will take to soften the dirt is found by trial; it could be a few minutes or days. On some surfaces the softened dirt can then be removed by hosing down the surface, but usually it is necessary to assist removal with the gentle use of bristle brushes and nonferrous or stainless-steel-wire brushes.

The low-pressure water spray method is effective only when the dirt lies lightly on the surface or is bound to the wall with water-soluble matter.

High-Pressure Water Blasting

With the recent development of ultra-high-pressure water-jetting equipment, water can be used to clean masonry surfaces effectively. High-pressure water blasting relies on the force of the water rather than on abrasives. Pressures up to 55,000 psi are available; however, most of the work is accomplished at 5,000 to 10,000 psi or less. Although usually not needed, sand can be injected into the high-pressure water stream to enhance cutting. Oils and grease are usually removed before water blasting.

A variety of equipment is available for this type of surface cleaning. Nozzles range from flat-fan pattern tips to a straight jet tip. The fan pattern acts as a blade that pries up and lifts away the undesirable surface accumulation. The straight jet could cut a hole completely through concrete masonry. The techniques used are similar to sandblasting: correct distance from the surface, nozzle angle, and pressure are determined by the type and amount of material to be removed. Water blasting can be used to prepare surfaces for coating, remove dirt or stains, or abrade the surface for repairs.

Graffiti Removal

A large number of commercially available products are suitable for removing spray-paint and felt-tip markings from concrete surfaces. These products are generally effective also for removing crayon, chalk, and lipstick. The manufacturer's directions should always be followed. If satisfactory results are not obtained with the first remover applied, a second or third attempt with other products should be made. A single product may not remove both spray-paint and felt-tip-pen stains.

If a proprietary cleaner is not available, methylene chloride can be used. While wearing protective clothing, brush methylene chloride onto the surface, wait 2 minutes, and rinse with water during continued brushing. Oxalic acid or hydrogen peroxide can be used to help bleach out some of the pigment from pores in the concrete masonry. Solutions of sodium hydroxide, xylene, or methyl ethyl ketone are also helpful in removing graffiti. Effective cleaning can also be accomplished with waterblasting and sandblasting.

After the graffiti is removed, or preferably before a structure is placed in service, an antigraffiti coating or sealer should be applied. The surface treatment should prevent graffiti from entering the pores of the masonry and should facilitate removal of the graffiti, preferably without removing the surface treatment.

Aliphatic urethanes are considered the best antigraffiti coatings because of the resistance to solvents, yellowing, and abrasion. Solvents such as mineral spirits or methyl ethyl ketone can remove most graffiti from an aliphatic polyurethane without compromising the urethane coating. Acrylics, epoxies, silanes, and siloxanes are also used to make graffiti removal easier; however, acrylics dissolve with the solvent and epoxies tend to yellow or discolor. Silanes and siloxanes may not resist certain graffiti materials as well as the urethanes, but they do maintain a high breathability at the surface while resisting penetration of graffiti materials into the pores of the masonry.

Dirt Removal

Airborne dirt can collect on any concrete masonry surface to form a dark and sometimes oily buildup or stain. Buildings may need to be cleaned of air pollution-induced dirt deposits to regain their original appearance. Some dirt can be removed by scrubbing with detergent and water or 1 part hydrochloric acid in about 20 parts water. However, special proprietary cleaners, made to remove dirt with minimal attack of the masonry, are often preferred over hydrochloric acid solutions that attack the cement paste in concrete masonry.

A solution of 1 part phosphoric acid to about 3 parts water can be used to scrub away light to moderate amounts of dirt with little to no attack of the surface.

Proprietary cleaners, made with hydrochloric acid and buffers to protect the concrete, are used to remove severe dirt buildup. An alkaline prewash followed by an acetic acid wash is another cleaning method. Special cleaning solutions can be specially designed to remove particular types of dirt.

The methods used to remove oil can be helpful in removing very oily dirt. Steam cleaning and light sandblasting or waterblasting are also effective as discussed earlier.

Once a surface is clean, it is good practice to apply a breathable clear sealer (such as a methacrylate or acrylic-based material) or a clear water-repellent penetrating sealer (such as silane or siloxane) to resist dirt buildup and make future cleaning easier. Some cleaning specialists prefer the silane or siloxane treatments for their high breathability (often with a 95% vapor transmission). Also see "Graffiti Removal."

Stain Removal

Concrete masonry can be stained by many substances. The first step in stain removal is to determine what caused the stain. Water-based or water-soluble stains, such as many foods, beverages, and soil, can be removed with water cleaning methods as discussed earlier. A detergent, water, and a stiff-bristle brush can be very effective in removing common stains. The cleaning techniques presented under "Cleaning Concrete Masonry Surfaces" are also applicable to removing stains. For more information on removing specific stains, refer to PCA's *Removing Stains and Cleaning Concrete Surfaces,* IS214T. This publication discusses the removal of common stains ranging from aluminum and food stains to oil, smoke, and wood stains. Discussions on removing efflorescence, dirt, and graffiti were presented earlier in this chapter.

Covering Stains

After cleaning a concrete masonry surface, some stains and discoloration may not come out as desired. A shadow of the stain may remain. Some stains or discoloration may be so severe or extensive it may not be economical or possible to remove them adequately. In these cases, it may be necessary to coat the concrete and cover up the stains to restore a pleasant uniform surface appearance. Before using a coating, be sure it can breathe as required, will properly adhere to the concrete masonry, is durable under the exposure conditions, will not discolor or fade, and will not allow the stain to bleed through. Stain bleeding can be easily tested by coating an inconspicuous area and observing it for bleeding over a few days. Because of the bleeding ability of some stains, an impermeable sealer or pretreatment may need to be applied prior to the coating application. Most coatings require some surface preparation (cleaning or roughening) to bond properly to the substrate.

Coatings (colored or white) commonly consist of polymer-modified cementitious products, silicone emulsions, and other materials. The concrete must also be sufficiently dry for certain coatings to be applied (see ASTM D4263). Many coatings also facilitate future surface cleanup. Special staining materials, such as penetrating acrylic resin and pigment stains, may be used in an attempt to color the surface and blend in the discloration to produce a more uniform appearance.

Cement plaster can be applied over vertical or underside surfaces to hide unsightly areas.

Protective Surface Treatments

After the masonry is cleaned, a clear (or colored) surface treatment can be applied to prevent future staining materials from penetrating the concrete, to facilitate stain or dirt removal, and to repel water from the surface to reduce water permeation. The sealer should be durable in the exposure conditions, properly adhere to or penetrate the concrete masonry surface, not discolor or yellow, and have appropriate breathing properties. Many sealers need a surface pretreatment. Most clear coatings or penetrating sealers enhance or do not affect the masonry appearance.

Materials used to protect masonry surfaces include aliphatic urethane, methyl methacrylate, various modified acrylics, and epoxy sealers as well as penetrating-water-repellent sealers such as silane and siloxane. The silanes and siloxanes allow a high breathability or vapor transmission, which is important to concrete masonry structures. Where colored surface treatments are needed, acrylics and styrene butadiene perform well. Asphaltic cement and other nonbreathable coatings should not be used unless all moisture in the masonry is removed. Moisture trapped in the masonry by the asphalt can cause deterioration of units and rusting of metal components. For locations in which a chemical is attacking the concrete masonry, an appropriate sealer that can resist the environment must be used (see PCA's *Effects of Substances on Concrete and Guide to Protective Treatments,* IS001T).

REFERENCES ON CONCRETE MASONRY

Since this handbook points out only the most essential facts concerning concrete masonry and its constituent materials, references have been made to many of the following publications to indicate the sources of material presented in the text. This list of references is also intended to serve as a guide for further study.

General References

1. *Masonry: Past and Present,* STP 589, American Society for Testing and Materials, Philadelphia, 1975.

2. *Menzel Symposium on High Pressure Steam Curing,* SP-32, American Concrete Institute, Detroit, 1972.

3. Menzel, Carl A., *General Considerations of Cracking in Concrete Masonry Walls and Means for Minimizing It,* Development Department Bulletin DX020, Portland Cement Association, 1958.

4. Copeland, R. E., and Carlson, C. C., "Tests of the Resistance to Rain Penetration of Walls Built of Masonry and Concrete," *Journal of the American Concrete Institute,* Proceedings Vol. 36, Detroit, November 1939, pages 169-192.

5. Hedstrom, R. O., *Load Tests of Patterned Masonry Walls,* Development Department Bulletin DX041, Portland Cement Association, 1961.

6. Hedstrom, R. O., *Tensile Testing of Concrete Block and Wall Elements,* Development Department Bulletin DX105, Portland Cement Association, 1966.

7. Kuenning, W. H., *Improved Method of Testing Tensile Bond Strength of Masonry Mortars,* Research Department Bulletin RX195, Portland Cement Association, 1966.

8. Hedstrom, R. O.; Litvin, Albert; and Hanson, J. A., *Influence of Mortar and Block Properties on Shrinkage Cracking of Masonry Walls,* Development Department Bulletin DX131, Portland Cement Association, 1968.

9. *Concrete Masonry Screen Walls,* NCMA-TEK 5, National Concrete Masonry Association, Herndon, Virginia, 1970.

10. *Concrete Masonry Cantilever Retaining Walls,* NCMA-TEK 4B, National Concrete Masonry Association, Herndon, Virginia, 1983.

11. Northwood, T. D., and Monk, D. W., *Sound Transmission Loss of Masonry Walls: Twelve-Inch Lightweight Concrete Blocks with Various Surface Finishes,* Building Research Note No. 90, National Research Council of Canada, Ottawa, April 1974.

12. *Nonreinforced Concrete Masonry Design Tables,* National Concrete Masonry Association, Herndon, Virginia, 1989.

13. Amrhein, James E., *Masonry Design Manual,* Masonry Institute of America (formerly Masonry Research), Los Angeles, 1969.

14. Amrhein, James E., *Reinforced Masonry Engineering Handbook,* Fourth Edition, Masonry Institute of America (formerly Masonry Research), Los Angeles, 1983.

15. *Residential Fireplace and Chimney Handbook,* Masonry Institute of America (formerly Masonry Research), Los Angeles, 1970.

16. *All-Weather Masonry Construction—State-of-the-Art Report,* Technical Task Committee Report, International Masonry Industry All-Weather Committee, December 1968.

17. Sears, Bradford G., *Retaining Walls,* American Society of Landscape Architects Foundation, Herndon, Virginia, 1973.

18. *Portland Cement Plaster (Stucco) Manual,* EB049M, Portland Cement Association, 1980.

19. Fishburn, Cyrus C., *Strength and Resistance to Corrosion of Ties for Cavity Walls,* BMS Report 101, National Bureau of Standards (now called National Institute of Standards and Technology), U.S. Department of Commerce, Washington, D.C., July 1943.

20. *Reinforced Concrete Masonry Design Tables,* National Concrete Masonry Association, Herndon, Virginia, 1971.

21. *Design of Concrete Masonry Warehouse Walls,* NCMA-TEK 37, National Concrete Masonry Association, Herndon, Virginia 1972.

22. Valore, Rudolph C., "Calculation of U-Valves of Hollow Concrete Masonry," *Concrete International,* American Concrete Institute, Detroit, February 1980, pages 40-63.

23. *Apartments—Design for Economy, Noise Control, and Fire Safety,* NCMA-TEK 51, National Concrete Masonry Association, Herndon, Virginia, 1973.

24. Leba, Theodore, Jr., *Design Manual, The Application of Non-Reinforced Concrete Masonry Load-Bearing Walls in Multi-Storied Structures,* National Concrete Masonry Association, Herndon, Virginia, 1969.

25. Mackintosh, Albyn, *Design Manual, The Application of Reinforced Concrete Masonry Load-Bearing Walls in Multi-Storied Structures,* National Concrete Masonry Association, Herndon, Virginia, 1968.

26. Isberner, Albert W., *Specifications and Selection of Materials for Masonry Mortars and Grouts,* Research and Development Bulletin RD024M, Portland Cement Association, 1974.

27. Isberner, Albert W., *Properties of Masonry Cement Mortars,* Research and Development Bulletin RD019M, Portland Cement Association, 1974.

28. *Acoustics of Concrete in Buildings,* IS159T, Portland Cement Association, 1982.

29. *ASHRAE Handbook of Fundamentals,* American Society of Heating, Refrigerating and Air-Conditioning Engineers, Inc., New York, 1990.

30. Brewer, Harold W., *General Relation of Heat Flow Factors to the Unit Weight of Concrete,* Development Department Bulletin DX114, Portland Cement Association, 1967.

31. Peavy, B. A.; Powell, F. J.,; and Burch, D. M., *Dynamic Thermal Performance of an Experimental Masonry Building,* Building Science Series No. 45, National Bureau of Standards (now called National Institute of Standards and Technology), U.S. Department of Commerce, Washington, D.C., 1973.

32. *Concrete Energy Conservation Guidelines,* EB083B, Portland Cement Association, 1980.

33. Harris, C. M., *Handbook of Noise Control,* McGraw-Hill Book Co., New York, 1957.

34. Litvin, Albert, and Belliston, Harold W., "Sound Transmission Loss Through Concrete and Concrete Masonry Walls," *Journal of the American Concrete Institute,* Detroit, December 1978. Also available from the Portland Cement Association as RD066M.

35. *Noise Abatement and Control,* Highway Research Record No. 448, Transportation Research Board (formerly the Highway Research Board), Washington, D.C., 1973.

36. Menzel, C.A., *Tests of the Fire Resistance and Strength of Walls of Concrete Masonry Units,* Portland Cement Association, January 1934.

37. Catani, Mario J., and Goodwin, Stanley E., "Heavy Building Envelopes and Dynamic Thermal Response," *Journal of the American Concrete Institute,* Proceedings Vol. 73, Detroit, February 1976, pages 83-86.

38. *Tall Concrete Masonry Walls,* NCMA-TEK103, National Concrete Masonry Association, Herndon, Virginia, 1978.

39. *Building Materials List,* Underwriters' Laboratories, Inc., Chicago, January 1974, pages 440-442.

40. Yokel, F. Y.; Mathey, R. G.; and Dikkers, R. D., *Compressive Strength of Slender Concrete Masonry Walls,* Building Science Series No. 33, National Bureau of Standards (now called National Institute of Standards and Technology), U.S. Department of Commerce, Washington, D.C., 1970.

41. Olin, H. B.; Schmidt, J. L.; and Lewis, W. H., *Construction Principles, Materials & Methods,* Fifth Edition, U.S. League of Savings Institutions, Chicago, 1983.

42. Yokel, Felix Y., "Engineers Inquiry Box," *Civil Engineering,* American Society of Civil Engineers, New York, December 1974, page 60.

43. *Efflorescence,* IS020T, Portland Cement Assocation, 1968.

44. Holm, Thomas A., "Engineered Masonry with High Strength Lightweight Concrete Masonry Units," *Concrete Facts,* Vol. 17, No. 2, Expanded Shale, Clay and Slate Institute, Salt Lake City, 1972, pages 9-16.

45. Litvin, Albert, *Clear Coatings for Exposed Architectural Concrete,* Development Department Bulletin DX137, Portland Cement Association, 1968.

46. Dickey, Walter L., "Reinforced Concrete Masonry Construction," *Handbook of Concrete Engineering,* Second Edition, ed. Mark Fintel, Van Nostrand Reinhold Company, New York, 1985, pages 632-662.

47. *Fire Resistance Index,* Underwriters' Laboratories, Inc., Chicago, January 1975.

48. *Fire Resistance of Expanded Shale, Clay and Slate Concrete Masonry,* Lightweight Concrete Information Sheet No. 14, Expanded Shale, Clay and Slate Institute, Salt Lake City, October 1971.

49. *Thermal Insulation of Various Walls,* Expanded Shale, Clay and Slate Institute, Salt Lake City, July 1972.

50. Diehl, John R., *Manual of Lathing and Plastering,* Mac Publishers Association, New York, 1960 (out of print).

51. Portland Cement Association, *Special Concretes, Mortars, and Products,* John Wiley & Sons, Inc., New York, 1975 (out of print), pages 269-342.

52. ACI Committee 531, "Concrete Masonry Structures—Design and Construction," *Journal of the American Concrete Institute,* Detroit, Proceedings Vol. 67, No. 5, May 1970, pages 380-403, and No. 6, June 1970, pages 442-460.

53. ACI Committee 531, "Proposed Standard Specifications for Concrete Masonry," *Journal of the American Concrete Institute,* Proceedings Vol. 72, No. 11, Detroit, November 1975, pages 614-627.

54. *Thermal Insulation of Concrete Masonry Walls,* NCMA-TEK 38-A, National Concrete Masonry Association, Herndon, Virginia, 1980.

55. *Tables of U-Values for Concrete Masonry Walls,* NCMA-TEK 67, National Concrete Masonry Association, Herndon, Virginia, 1975.

56. *Building Code Requirements for Reinforced Masonry (A41.2),* National Bureau of Standards Handbook 74, American National Standards Institute, New York, 1960.

57. *American Standard Building Code Requirements for Masonry (A41.1),* National Bureau of Standards Misc. Publication 211, American National Standards Institute, New York, 1954.

58. *Specification for the Design and Construction of Load-Bearing Concrete Masonry,* TR75-B, National Concrete Masonry Association, Herndon, Virginia, 1968.

59. *Masonry Structural Design for Buildings,* Army Technical Manual TM5-809-3, or Air Force Manual AFM 88-3, Chapter 3, Departments of the Army and Air Force, Washington, D.C., 1973.

60. *Seismic Design for Buildings,* TM5-809-10, Department of the Army, or AFM 88-3, Chapter 13, Department of the Air Force, or NAVDOCKS P-355, Department of the Navy, Washington, D.C., 1966.

61. *Fire-Resistance Classifications of Building Construction,* BMS Report 92, National Bureau of Standards (now called National Institute of Standards and Technology), U.S. Department of Commerce, Washington, D.C., 1942.

62. *Recommended Practices & Guide Specifications for Cold Weather Masonry Constuction,* International Masonry Industry All-Weather Council, 1970. Available from Portland Cement Association as LT107M.

63. *Guide Specification for Military and Civil Works Construction—Masonry,* CE-206.01, Corps of Engineers, Department of the Army, Washington, D.C., 1968.

64. *Concrete Masonry Basement Walls,* NCMA-TEK 56A, National Concrete Masonry Association, Herndon, Virignia, 1982.

65. *Guide Sepcification for Concrete Masonry,* National Concrete Masonry Association, Herndon, Virginia, 1971.

66. *Concrete Masonry Units,* Seventh Edition, UL618, Underwriters' Laboratories, Inc., Chicago, 1975.

67. *CABO One and Two Family Dwelling Code,* 1989 Edition, The Council of American Building Officials, Falls Church, Virginia.

68. *Minimum Design Loads for Buildings and Other Structures,* ASCE 7-88 (Revision of ANSI A58.1-82), American Society of Civil Engineers, New York, 1990.

69. *Fire Protection Handbook,* Thirteenth Edition, National Fire Protection Assocation, Boston, 1969.

70. *Accelerated Curing of Concrete at Atmospheric Pressure,* ACI 517.2R-87, American Concrete Institute, Detroit, 1987.

71. Dubovoy, Val S., *Evaluation of Selected Properties of Masonry Mortars—Special Testing Program,* Portland Cement Association, 1990.

72. *Footer Block Detail Handbook,* IDR 4016, National Concrete Masonry Association, Herndon, Virginia, 1987.

73. Amrhein, James R., *Informational Guide to Grouting Masonry,* Masonry Institute of America, Los Angeles, 1990.

74. Amrhein, James E., and Merrigan, Michael W., *Reinforced Concrete Masonry Construction Inspector's Handbook,* Masonry Institute of America, Los Angeles, and International Conference of Building Officials, Whittier, California, 1989.

75. *Building Code Requirements for Masonry Structures,* ACI 530-88/ASCE 5-88, American Concrete Institute, Detroit, and American Society of Civil Engineers, New York, 1988.

76. *Specifications for Masonry Structures,* ACI 530.1-88/ASCE 6-88, American Concrete Institute, Detroit, and American Society of Civil Engineers, New York, 1988.

77. *Commentary on Building Code Requirements for Masonry Structures,* ACI 530-88/ASCE 5-88, and *Commentary on Specifications for Masonry Structures,* ACI 530.1-88/ASCE 6-88, American Concrete Institute, Detroit, and American Society of Civil Engineers, New York, 1988.

78. Laska, Walter, "Detailing Brick Spandrels," *Magazine of Masonry Construction,* The Aberdeen Group, Addison, Illinois, June 1990, pages 252-254.

79. *The BOCA National Building Code,* Eleventh Edition, Building Officials and Code Administrators International, Inc., Country Club Hills, Illinois, 1990.

80. *Standard Building Code,* Southern Building Code Congress International, Birmingham, Alabama, 1988.

81. *Uniform Building Code,* International Conference of Building Officials, Whittier, California, 1988.

82. *Analytical Methods of Determining Fire Endurance of Concrete and Masonry Members—Model Code Approved Procedures,* SR267B, Concrete and Masonry Industry Firesafety Committee, Skokie, Illinois, 1985.

83. *Steel Column Fire Protection,* NCMA-TEK 128, National Concrete Masonry Association, Herndon, Virginia, 1983.

84. *Standard for Chimneys, Fireplaces, Vents, and Solid Fuel Burning Appliances,* NFPA 211, National Fire Protection Association, Quincy, Massachusetts, 1984.

85. *Cultured Stone,* Stucco Stone Products, Inc., Napa, California, 1989.

86. *Building Movements and Joints,* EB086B, Portland Cement Association, 1982.

87. Heslip, John A., "Masonry/Metal Stud Wall Systems, A Critical Review—Updated," *The Story Pole,* Vol. 14, No. 2, Masonry Institute of Michigan, Inc., Farmington, Michigan, April 1983.

88. "Brick Veneer Steel Stud Panel Walls," *Technical Notes on Brick Construction,* No. 28B Revised II, Brick Institute of America, Reston, Virginia, February 1987.

89. "Structural Backup Systems for Concrete Masonry Veneers," NCMA-TEK 114A, National Concrete Masonry Association, Herndon, Virginia, 1981.

90. Piper, Richard, "Troubles with Synthetic Stucco," *New England Builder,* Builderburg Partners, Ltd., Washington, D.C., June 1988, pages 34-38.

91. Foster, Allan, "Performance Characteristics of Exterior Plastering Systems," *The Construction Specifier,* Construction Specifications Institute, Alexandria, Virginia, August 1989, pages 84-93.

92. Labs, Kenneth, "P/A Technics, Inside EIF Systems," *Progressive Architecture,* Reinhold Publishing, Cleveland, October 1989, pages 100-106.

93. *Segmental Masonry Footings,* NCMA-TEK 110A, National Concrete Masonry Association, Herndon, Virginia, 1988.

94. *Laboratory Research on Water Permeance of Masonry,* prepared for the Masonry Cement Research and Education Group, Portland Cement Association, 1980.

95. *The Effect of Outdoor Exposure on Water Permeance of Masonry,* prepared for the Masonry Cement Research and Education Group, Portland Cement Association, 1982.

96. Isberner, A. W., *Water Permeance of Masonry,* LT111M, Portland Cement Association, 1979.

97. Schneider, Robert R., and Dickey, Walter L., *Reinforced Masonry Design, Second Edition,* Prentice-Hall, Inc., Englewood Cliffs, New Jersey, 1987.

98. *Strength Design—A New Engineering Guide for Concrete Masonry Walls,* First Edition, Concrete Masonry Association of California and Nevada, Citrus Heights, California, 1988.

99. *Concrete Masonry Lintels,* NCMA-TEK 25A, National Concrete Masonry Association, Herndon, Virginia, 1985.

100. *Noise Control with Concrete Masonry in Multi-Family Housing,* NCMA-TEK 18A, National Concrete Masonry Association, Herndon, Virginia, 1983.

American Society for Testing and Materials (ASTM)*

C5	Specification for Quicklime for Structural Purposes
C31	Method of Making and Curing Concrete Test Specimens in the Field
C33	Specification for Concrete Aggregates
C39	Method of Test for Compressive Strength of Cylindrical Concrete Specimens
C55	Specification for Concrete Building Brick
C90	Specification for Hollow Load-Bearing Concrete Masonry Units
C91	Specification for Masonry Cement
C94	Specification for Ready-Mixed Concrete
C129	Specification for Non-Load-Bearing Concrete Masonry Units
C139	Specification for Concrete Masonry Units for Construction of Catch Basins and Manholes
C140	Methods of Sampling and Testing Concrete Masonry Units
C143	Test Method for Slump of Hydraulic Cement Concrete
C144	Specification for Aggregate for Masonry Mortar
C145	Specification for Solid Load-Bearing Concrete Masonry Units
C150	Specification for Portland Cement
C207	Specification for Hydrated Lime for Masonry Purposes
C236	Test Method for Steady-State Thermal Performance of Building Assemblies by Means of a Guarded Hot Box
C270	Specification for Mortar for Unit Masonry
C315	Specification for Clay Flue Linings
C404	Specification for Aggregates for Masonry Grout
C423	Test Method for Sound Absorption and Sound Absorption Coefficients by the Reverberation Room Method
C426	Test Method for Drying Shrinkage of Concrete Block
C476	Specification for Grout for Masonry
C595	Specification for Blended Hydraulic Cements
C617	Practice for Capping Cylindrical Concrete Specimens
C631	Specification for Bonding Compounds for Interior Plastering
C634	Definitions of Terms Relating to Environmental Acoustics
C744	Specification for Prefaced Concrete and Calcium Silicate Masonry Units
C780	Method for Pre-Construction and Construction Evaluation of Mortars for Plain and Reinforced Unit Masonry
C847	Specification for Metal Lath
C887	Specification for Packaged, Dry, Combined Materials for Surface Bonding Mortar
C897	Specification for Aggregate for Job-Mixed Portland Cement-Based Plasters
C901	Specification for Prefabricated Masonry Panels
C920	Specification for Elastomeric Joint Sealants
C926	Specification for Application of Portland Cement-Based Plaster
C932	Specification for Surface-Applied Bonding Agents for Exterior Plastering

*ASTM, 1916 Race Street, Philadelphia, Pa. 19103,
Phone 215-299-5400.

C933	Specification for Welded Wire Lath
C936	Specification for Solid Concrete Interlocking Paving Units
C946	Practice for Construction of Dry-Stacked, Surface-Bonded Walls
C952	Test Method for Bond Strength of Mortar to Masonry Units
C962	Guide for Use of Elastomeric Joint Sealants
C1006	Test Method for Splitting Tensile Strength of Masonry Units
C1012	Test Method for Length Change of Hydraulic-Cement Mortars Exposed to a Sulfate Solution
C1019	Method of Sampling and Testing Grout
C1032	Specification for Woven Wire Plaster Base
C1038	Test Method for Expansion of Portland Cement Mortar Bars Stored in Water
C1063	Specification for Installation of Lathing and Furring for Portland Cement-Based Plaster
C1072	Method for Measurement of Masonry Flexural Bond Strength
C1093	Practice for the Accreditation of Testing Agencies for Unit Masonry
C1142	Specification for Ready Mixed Mortar for Unit Masonry
C1148	Test Method for Measuring the Drying Shrinkage of Masonry Mortar
D3665	Practice for Random Sampling of Construction Materials
D4258	Practice for Surface Cleaning Concrete for Coating
D4259	Practice for Abrading Concrete
D4260	Practice for Acid Etching Concrete
D4261	Practice for Surface Cleaning Concrete Unit Masonry for Coating
D4262	Test Method for pH of Chemically Cleaned or Etched Concrete Surfaces
D4263	Test Method for Indicating Moisture in Concrete by the Plastic Sheet Method
E72	Methods of Conducting Strength Tests of Panels for Building Construction
E90	Method for Laboratory Measurement of Airborne Sound Transmission Loss of Building Partitions
E119	Methods of Fire Tests of Building Construction and Materials
E336	Method for Measurement of Airborne Sound Insulation in Buildings
E380	Practice for Use of the International System of Units (SI) (the Modernized Metric System)
E413	Classification for Determination of Sound Transmission Class
E447	Test Methods for Compressive Strength of Masonry Prisms
E492	Method of Laboratory Measurement of Impact Sound Transmission Through Floor-Ceiling Assemblies Using the Tapping Machine
E597	Practice for Determining a Single-Number Rating of Airborne Sound Isolation in Multi-Unit Building Specifications

Appendix A
DETAILS OF CONCRETE MASONRY CONSTRUCTION

On the following pages are a number of details that might be encountered in buildings using concrete masonry. These details are offered only as a *guide* for design and are not guaranteed for completeness or suitability for all buildings in all places. For example, although assumed to be present, weepholes are not always illustrated with flashing details. Building designs involve a wide range of shapes, dimensions, materials, loads, uses, and climates. While some simple details are adequate for many ordinary structures, there are circumstances that require they be refined or improved, as in regions with earthquakes or high winds. The knowledgeable and experienced designer or builder will recognize the applications of various configurations and be able to refine or improve these details to fit the individual project.

The connection of floors and roofs to masonry walls deserves particular attention. Some building codes, particularly where earthquakes are common, require a positive connection between the floor or roof and the masonry wall.* Although practice varies from one region to another, engineered concrete masonry structures most often have reinforcing bar connections between wall and floor or roof. Wooden framing must always be anchored. In some cases the dead-load friction of a concrete or steel deck on the wall will suffice. The following friction coefficients, which are based on a safety factor of 2, are suggested:

Connection	Friction coefficient
Steel to steel	0.12
Cast-in-place concrete to steel	0.20

Cast-in-place concrete to hardboard 0.25
Cast-in-place concrete to cast-in-
    place concrete 0.40
Precast concrete to concrete masonry 0.40
Cast-in-place concrete to concrete
    masonry 0.50

Where a floor is embedded in a wall, the connection will be stronger than for a roof that is merely resting on top of the wall. A roof may be subject to temperature and shrinkage movements and, if it is to act as a diaphragm and the structure is to act as a unit, some positive connection between wall and roof may be required.

The importance of construction details is well known. Not only do details affect the initial construction cost, but also they have an important influence on the behavior of the building under the traffic of use and the influence of weather. Since the cost of repairs frequently outweighs the cost of construction done properly at the outset, the architect, engineer, and builder should plan all details carefully.

Floor plans and designs of masonry residential structures are available from Home Building Plan Service, 2235 NE Sandy Blvd., Portland, Oregon 97232-2884 (Phone: 503-234-9337). Commercial and industrial designs and details are available in publications from the National Concrete Masonry Association, P.O. Box 781, Herndon, Virginia 22070 (Phone: 703-435-4900).

---

*See Ref. 42.

Foundation Details

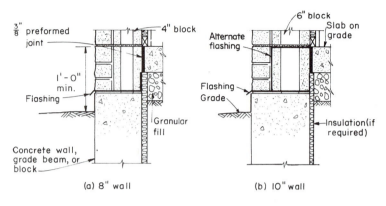

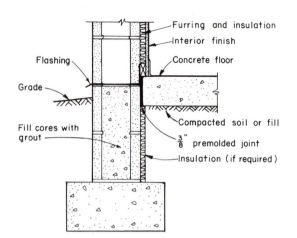

Fig. A-1. Foundation for composite wall.

Fig. A-4. Spread footing foundation.

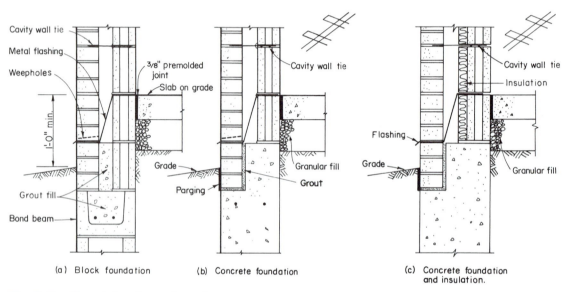

Fig. A-2. Foundation for cavity walls.

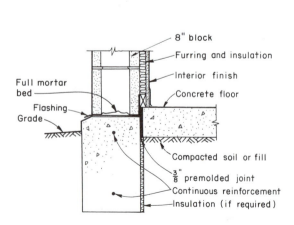

Fig. A-3. Trench-type foundation.

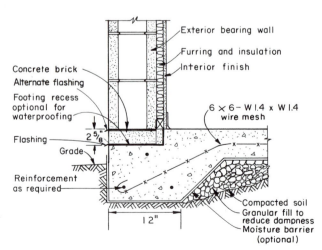

Fig. A-5. Floating slab foundation.

Basement Wall Details

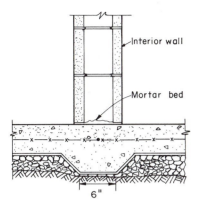

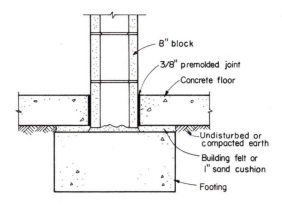

Fig. A-6. Load-bearing wall on slab.

Fig. A-9. Interior wall on footing.

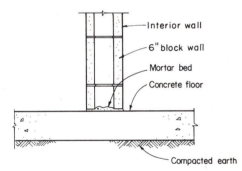

Fig. A-7. Non-load-bearing wall on slab.

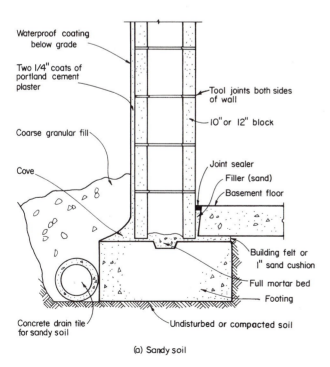

(a) Sandy soil

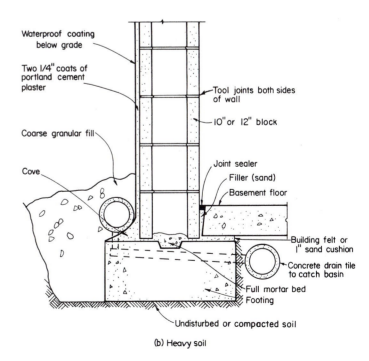

(b) Heavy soil

Fig. A-8. Exterior wall on footing.

Wall Details at Precast Concrete Floors

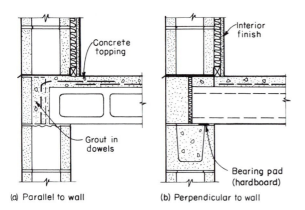

(a) Parallel to wall   (b) Perpendicular to wall

Fig. A-10. Hollow-core slab floors and single-wythe walls.

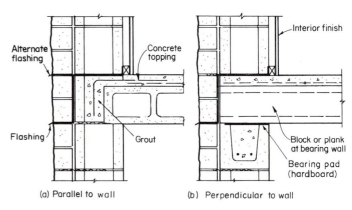

(a) Parallel to wall   (b) Perpendicular to wall

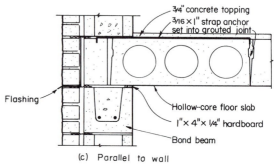

(c) Parallel to wall

Fig. A-13. Hollow-core slab floor and composite walls.

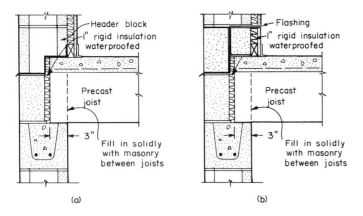

(a)   (b)

Fig. A-11. Precast joist floor and single-wythe wall.

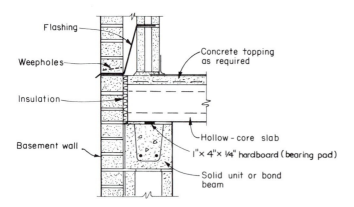

Fig. A-12. Hollow-core slab floors with cavity wall above.

Wall Details at Cast-in-Place Concrete Floors

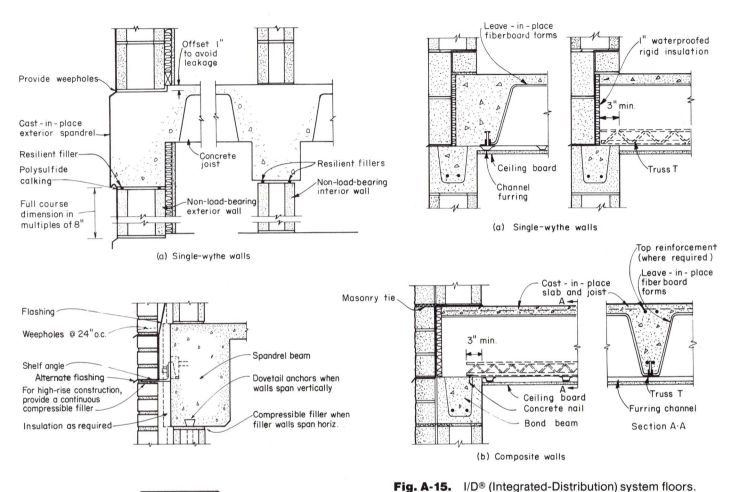

Offset 1" to avoid leakage

Provide weepholes

Cast-in-place exterior spandrel

Resilient filler

Polysulfide calking

Full course dimension in multiples of 8"

Concrete joist

Resilient fillers

Non-load-bearing interior wall

Non-load-bearing exterior wall

(a) Single-wythe walls

Leave-in-place fiberboard forms

1" waterproofed rigid insulation

3" min.

Ceiling board

Channel furring

Truss T

(a) Single-wythe walls

Flashing

Weepholes @ 24" o.c.

Shelf angle

Alternate flashing

For high-rise construction, provide a continuous compressible filler

Insulation as required

Spandrel beam

Dovetail anchors when walls span vertically

Compressible filler when filler walls span horiz.

Masonry tie

Cast-in-place slab and joist

Top reinforcement (where required)

Leave-in-place fiber board forms

3" min.

Ceiling board

Concrete nail

Bond beam

Truss T

Furring channel

Section A·A

(b) Composite walls

Fig. A-15. I/D® (Integrated-Distribution) system floors.

®Trademark and servicemark of the Portland Cement Association 1971.

Concrete column

Continuous anchor slot

Metal ties at 16" o.c.

Control joint

Cavity wall tie

(b) Cavity walls

Fig. A-14. Curtain wall and partition with concrete frame.

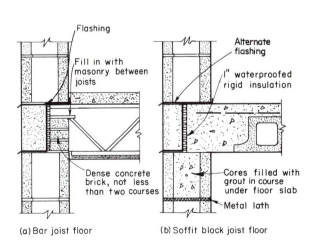

Flashing

Fill in with masonry between joists

Dense concrete brick, not less than two courses

Alternate flashing

1" waterproofed rigid insulation

Cores filled with grout in course under floor slab

Metal lath

(a) Bar joist floor

(b) Soffit block joist floor

Fig. A-16. Joist floors.

Wall Details at Concrete and Metal Roofs

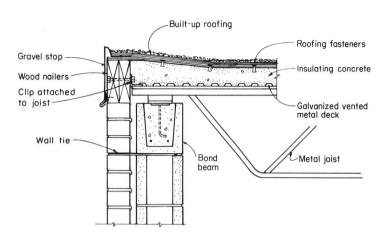

Fig. A-17. Metal deck roof.

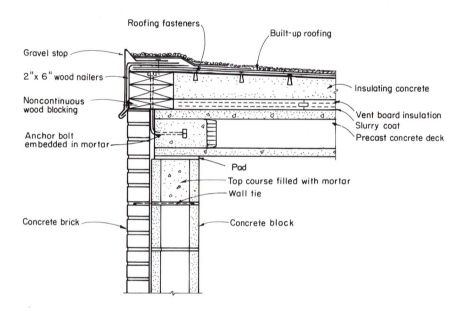

Fig. A-19. Precast concrete deck.

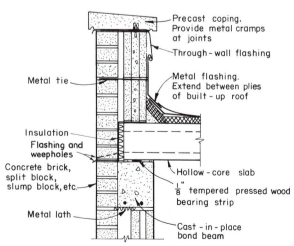

Fig. A-18. Parapet for load-bearing cavity walls.

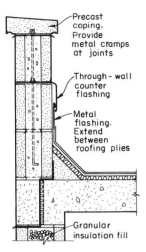

Fig. A-20. Parapet for flat concrete roof on single-wythe walls.

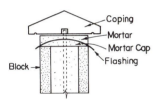

Fig. A-20a. Coping anchorage and flashing detail.

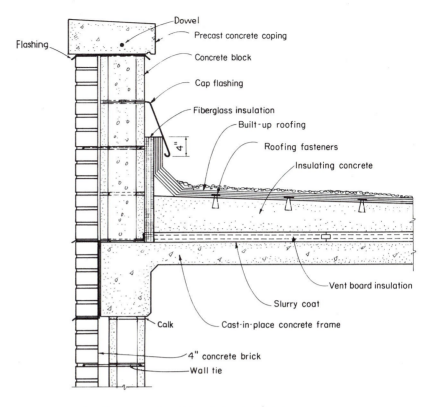

Fig. A-22. Parapet for concrete frame.

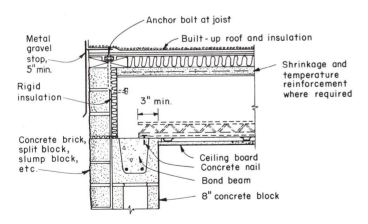

Fig. A-21. I/D® roof.

Wall Details at Wood Roofs and Floors

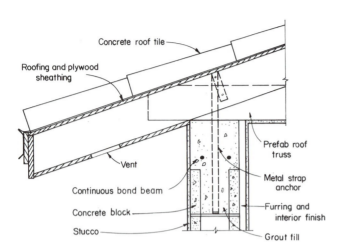

Fig. A-23. Low-sloped wood roofs with bond beam (hurricane-resistant).

Labels: Concrete roof tile; Roofing and plywood sheathing; Vent; Continuous bond beam; Concrete block; Stucco; Prefab roof truss; Metal strap anchor; Furring and interior finish; Grout fill

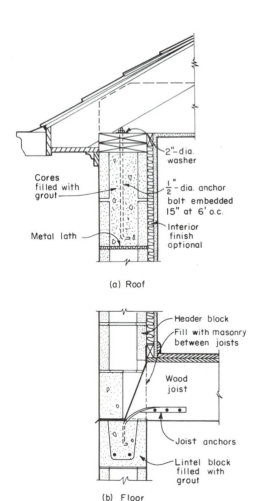

(a) Roof

Labels: 2"-dia. washer; Cores filled with grout; Metal lath; ½"-dia. anchor bolt embedded 15" at 6' o.c.; Interior finish optional

(b) Floor

Labels: Header block; Fill with masonry between joists; Wood joist; Joist anchors; Lintel block filled with grout

Fig. A-24. Steeply sloped wood roof with single-wythe walls.

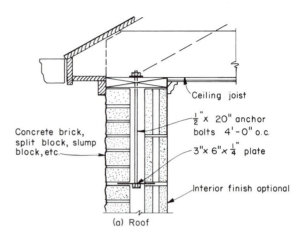

(a) Roof

Labels: Concrete brick, split block, slump block, etc.; Ceiling joist; ½" x 20" anchor bolts 4'-0" o.c.; 3"x 6"x ¼" plate; Interior finish optional

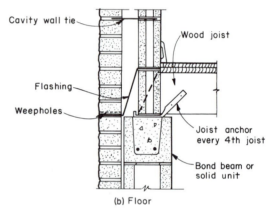

(b) Floor

Labels: Cavity wall tie; Flashing; Weepholes; Wood joist; Joist anchor every 4th joist; Bond beam or solid unit

Fig. A-25. Wood floors and roofs with cavity walls.

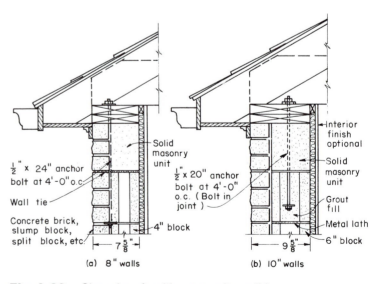

(a) 8" walls

Labels: ½" x 24" anchor bolt at 4'-0" o.c.; Wall tie; Concrete brick, slump block, split block, etc.; Solid masonry unit; 7 5/8"

(b) 10" walls

Labels: ½" x 20" anchor bolt at 4'-0" o.c. (Bolt in joint); 4" block; Interior finish optional; Solid masonry unit; Grout fill; Metal lath; 6" block; 9 5/8"

Fig. A-26. Sloped roofs with composite walls.

Door Frame Details

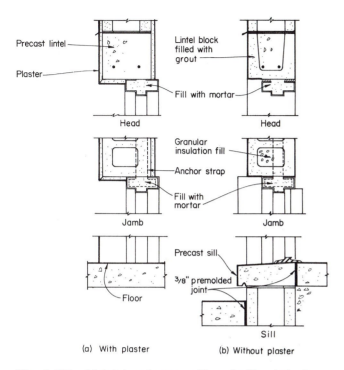

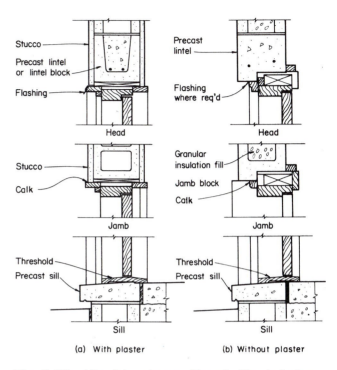

Fig. A-27. Metal door frames with and without plaster.

(a) With plaster   (b) Without plaster

Fig. A-27. Metal door frames with and without plaster.

(a) With plaster   (b) Without plaster

Fig. A-29. Wood door frame with and without plaster.

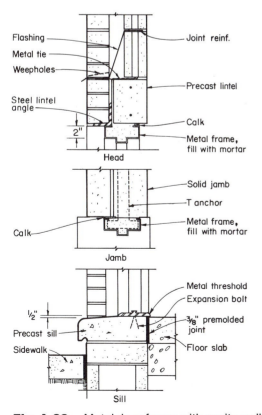

Fig. A-28. Metal door frame with cavity walls.

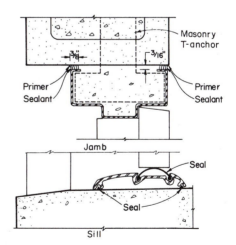

Fig. A-30. Details of metal door frames.

Metal Window Details

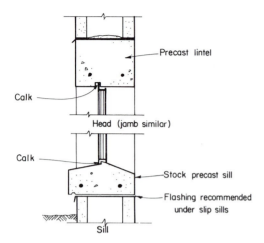

Fig. A-31. Metal basement window.

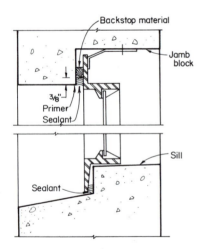

Fig. A-32. Metal window frame.

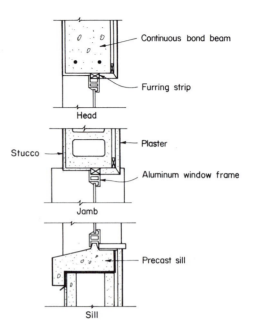

Fig. A-33. Aluminum window with plaster.

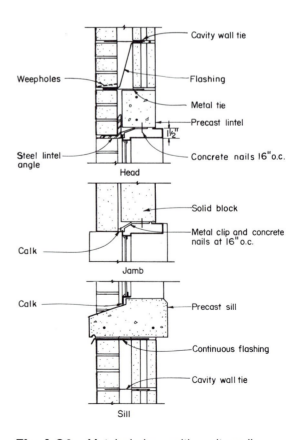

Fig. A-34. Metal windows with cavity walls.

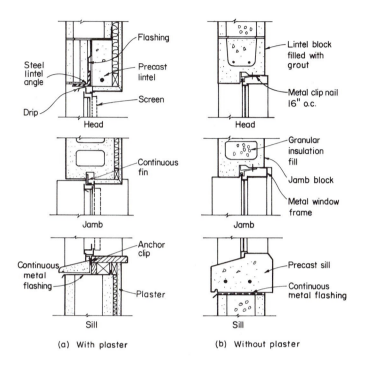

Fig. A-35. Metal windows with and without plaster.

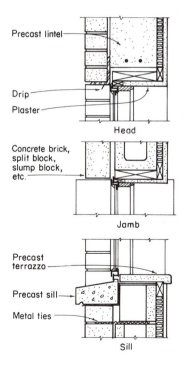

Fig. A-36. Metal windows for composite walls with plaster.

Wood Window Details

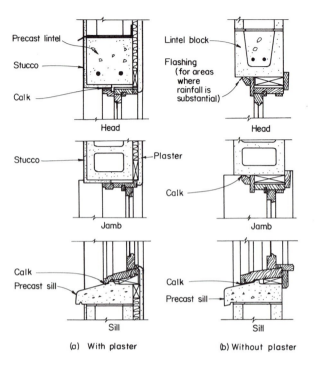

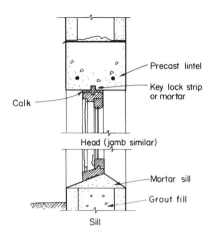

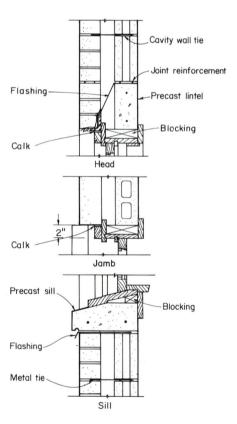

Fig. A-39. Basement wood windows.

Fig. A-37. Double-hung wood windows with and without plaster.

(a) With plaster   (b) Without plaster

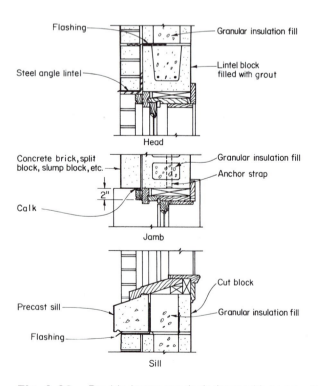

Fig. A-38. Double-hung wood windows with composite wall.

Fig. A-40. Wood windows with cavity wall.

Wall Details for Wood Framing

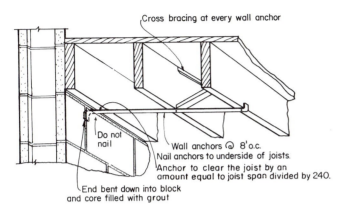

Cross bracing at every wall anchor

Do not nail

Wall anchors @ 8' o.c.
Nail anchors to underside of joists.
Anchor to clear the joist by an amount equal to joist span divided by 240.

End bent down into block and core filled with grout

Fig. A-41. Wood joists parallel to wall.

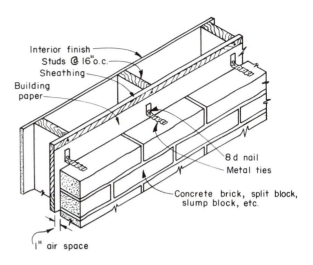

Interior finish
Studs @ 16" o.c.
Sheathing
Building paper

8 d nail
Metal ties

Concrete brick, split block, slump block, etc.

1" air space

Fig. A-43. Masonry veneer anchorage.

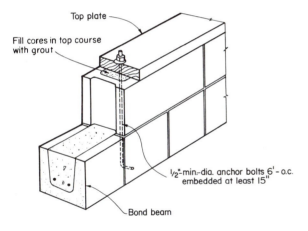

Top plate

Fill cores in top course with grout

½"-min.-dia. anchor bolts 6' - o.c. embedded at least 15"

Bond beam

Fig. A-42. Top plate anchorage over wall opening.

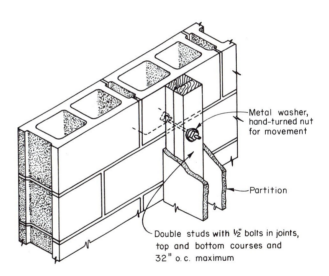

Metal washer, hand-turned nut for movement

Partition

Double studs with ½" bolts in joints, top and bottom courses and 32" o.c. maximum

Fig. A-44. Partition anchorage.

Appendix B
Metric Conversion Factors

The following list provides the conversion relationship between U.S. customary units and SI (International System) units. The proper conversion procedure is to multiply the specified value on the left (primarily U.S. customary values) by the conversion factor exactly as given below and then round to the appropriate number of significant digits desired. For example, to convert 11.4 ft. to meters: $11.4 \times 0.3048 = 3.47472$, which rounds to 3.47 meters. Do not round either value before performing the multiplication, as accuracy would be reduced. A complete guide to the SI system and its use can be found in ASTM E 380.

To convert from	to	multiply by	
Length			
inch (in.)	micron (μ)	25,400	E*
inch (in.)	centimeter (cm)	2.54	E
inch (in.)	meter (m)	0.0254	E
foot (ft)	meter (m)	0.3048	E
yard (yd)	meter (m)	0.9144	
Area			
square foot (sq ft)	square meter (sq m)	0.09290304	E
square inch (sq in.)	square centimeter (sq cm)	6.452	E
square inch (sq in.)	square meter (sq m)	0.00064516	E
square yard (sq yd)	square meter (sq m)	0.8361274	
Volume			
cubic inch (cu in.)	cubic centimeter (cu cm)	16.387064	
cubic inch (cu in.)	cubic meter (cu m)	0.00001639	
cubic foot (cu ft)	cubic meter (cu m)	0.02831685	
cubic yard (cu yd)	cubic meter (cu m)	0.7645549	
gallon (gal) Can. liquid	liter	4.546	
gallon (gal) Can. liquid	cubic meter (cu m)	0.004546	
gallon (gal) U.S. liquid**	liter	3.7854118	
gallon (gal) U.S. liquid	cubic meter (cu m)	0.00378541	
fluid ounce (fl oz)	milliliters (ml)	29.57353	
fluid ounce (fl oz)	cubic meter (cu m)	0.00002957	
Force			
kip (1000 lb)	kilogram (kg)	453.6	
kip (1000 lb)	newton (N)	4,448.222	
pound (lb) avoirdupois	kilogram (kg)	0.4535924	
pound (lb)	newton (N)	4.448222	
Pressure or stress			
kip per square inch (ksi)	megapascal (MPa)	6.894757	
kip per square inch (ksi)	kilogram per square centimeter (kg/sq cm)	70.31	
pound per square foot (psf)	kilogram per square meter (kq/sq m)	4.8824	
pound per square foot (psf)	pascal (Pa)†	47.88	
pound per square inch (psi)	kilogram per square centimeter (kg/sq cm)	0.07031	
pound per square inch (psi)	pascal (Pa)†	6,894.757	

pound per square inch (psi)	megapascal (MPa)	0.00689476	
Mass (weight)			
pound (lb) avoirdupois	kilogram (kg)	0.4535924	
ton, 2000 lb	kilogram (kg)	907.1848	
grain	kilogram (kg)	0.0000648	
Mass (weight) per length			
kip per linear foot (klf)	kilogram per meter (kg/m)	0.001488	
pound per linear foot (plf)	kilogram per meter (kg/m)	1.488	
Mass per volume (density)			
pound per cubic foot (pcf)	kilogram per cubic meter (kg/cu m)	16.01846	
pound per cubic yard (lb/cu yd)	kilogram per cubic meter (kg/cu m)	0.5933	
Temperature			
degree Fahrenheit (°F)	degree Celsius (°C)	$t_C = (t_F - 32)/1.8$	
degree Fahrenheit (°F)	degree Kelvin (°K)	$t_K = (t_F + 459.7)/1.8$	
degree Kelvin (°K)	degree Celsius (C°)	$t_C = t_K - 273.15$	
Energy and heat			
British thermal unit (Btu)	joule (J)	1055.056	
calorie (cal)	joule (J)	4.1868	E
Btu/°F · hr · ft²	W/m² · °K	5.678263	
kilowatt-hour (kwh)	joule (J)	3,600,000.	E
British thermal unit per pound (Btu/lb)	calories per gram (cal/g)	0.55556	
British thermal unit per hour (Btu/hr)	watt (W)	0.2930711	
Power			
horsepower (hp) (550 ft.lb/sec)	watt (W)	745.6999	E
Velocity			
mile per hour (mph)	kilometer per hour (km/hr)	1.60934	
mile per hour (mph)	meter per second (m/s)	0.44704	
Permeability			
darcy	centimeter per second (cm/sec)	0.000968	
feet per day (ft/day)	centimeter per second (cm/sec)	0.000352	

*E indicates that the factor given is exact.
**One U.S. gallon equals 0.8327 Canadian gallon.
†A pascal equals 1.000 newton per square meter.

Note:
One U.S. gallon of water weighs 8.34 pounds (U.S.) at 60°F and has a volume of 0.134 cu ft.
One cubic foot of water weighs 62.4 pounds (U.S.) and is 7.48 gallons.
One milliliter of water has a mass of 1 gram and has a volume of one cubic centimeter.
One U.S. bag of Portland cement weighs 94 lb.

INDEX

Related Materials

The following PCA materials may be of interest to readers of the **Concrete Masonry Handbook.**

Research and Development Bulletins

DX003 Investigation of the Moisture-Volume Stability of Concrete Masonry Units, by Joseph J. Shideler, 1955.

DX004 A Method for Determining the Moisture Condition of Hardened Concrete in Terms of Relative Humidity, by Carl A. Menzel, 1955.

DX013 Effect of Variations in Curing and Drying on the Physical Properties of Concrete Masonry Units, by William H. Kuenning and C. C. Carlson, 1956.

DX014 Lightweight Aggregates for Concrete Masonry Units, by C. C. Carlson, 1956.

DX020 General Considerations of Cracking in Concrete Masonry Walls and Means for Minimizing It, by Carl A. Menzel, 1958.

DX041 Load Tests of Patterned Masonry Walls, by R. O. Hedstrom, 1961.

DX064 Plant Drying and Carbonation of Concrete Block—NCMA-PCA Cooperative Program, by H. T. Toennies and J. J. Shideler, 1963.

DX069 Carbonation Shrinkage of Concrete Masonry Units, by J. J. Shideler, 1963.

DX105 Tensile Testing of Concrete Block and Wall Elements, by Richard O. Hedstrom, 1966.

DX114 General Relation of Heat Flow Factors to the Unit Weight of Concrete, by Harold W. Brewer, 1967.

DX131 Influence of Mortar and Block Properties on Shrinkage Cracking of Masonry Walls, by Richard O. Hedstrom, Albert Litvin, and J. A. Hanson, 1968.

RD019 Properties of Masonry Cement Mortars, by Albert W. Isberner, 1974.*

RD024 Specifications and Selection of Materials for Masonry Mortars and Grouts, by Albert W. Isberner, 1974.*

RD066 Sound Transmission Loss Through Concrete and Concrete Masonry Walls, by Albert Litvin and Harold W. Belliston, 1978.

RD067 Behavior of Inorganic Materials in Fire, by M. S. Abrams, 1979.

RD071 Thermal Performance of Masonry Walls, by A. E. Fiorato and C. R. Cruz, 1981.

RD075 Heat Transfer Characteristics of Walls Under Dynamic Temperature Conditions, by A. E. Fiorato, 1981.

RD095 Masonry Cement Mortars—A Laboratory Investigation, by V. S. Dubovoy and J. W. Ribar, 1990.

RX084 Fallacies in the Current Percent of Total Absorption Method for Determining and Limiting the Moisture Content of Concrete Block, by Carl A. Menzel, 1957.

RX195 Improved Method of Testing Tensile Bond Strength of Masonry Mortars, by W. H. Kuenning, 1966.

Engineering Bulletins

EB001T Design and Control of Concrete Mixtures

EB049M Portland Cement Plaster (Stucco) Manual

EB086B Building Movements and Joints

EB111T Cementitious Grouts and Grouting

Information Sheets and Pamphlets

IS001T Effects of Substances on Concrete and Guide to Protective Treatments

IS040M Mortars for Masonry Walls

IS159T Acoustics of Concrete in Buildings

IS181M Masonry Cement Mortars

---

*Available from the PCA library.

IS214T Removing Stains and Cleaning Concrete Surfaces
IS219M Permeability Tests of Masonry Walls
IS220M Building Weather-Resistant Masonry Walls
IS228P Structural Design of Pavements Surfaced with Concrete Paving Block (Pavers)
PA043M Recommended Practices for Laying Concrete Block

Special Publications, Special Reports, and Literature of Other Organizations

SP038H The Homeowner's Guide to Building with Concrete, Brick, and Stone
SR205B Luxury Apartment Developer Depends on Concrete Masonry and Prestressed Concrete
SR265B Choose Concrete and Masonry for Energy Savings
SR267B Analytical Methods of Determining Fire Endurance of Concrete and Masonry Members—Model Code-Approved Procedures
SR274B Discover the Benefits of Concrete and Masonry for Firesafety
SR284B Homebuilders Enhance Sales with Interlocking Concrete Pavers
SR291B Guide to BOCA/NBC Requirements for Concrete and Masonry Fire Walls
SR292B Guide to SBC Requirements for Concrete and Masonry Fire Walls
SR293B Guide to UBC Requirements for Concrete and Masonry Area Separation Walls
LT137D Handbook of Concrete Engineering
LT141M Quality Concrete Brick
LT160M Plaster and Drywall Systems Manual

Slide Sets

CS008 Building with Concrete Brick Veneers
SS018 Quality Concrete Masonry Construction Details
SS021 Control Joints for Concrete Masonry

To order, write or call Order Processing, Portland Cement Association, 5420 Old Orchard Road, Skokie, Illinois 60077-1083. Phone 708-966-9559 or 708-966-6200.

A complete list of PCA materials is available in the Association's free catalog of publications, computer software, and audiovisual materials (MS254G).